OECD Studies on Water

Drying Wells, Rising Stakes

TOWARDS SUSTAINABLE AGRICULTURAL GROUNDWATER USE

OECD
BETTER POLICIES FOR BETTER LIVES

This work is published under the responsibility of the Secretary-General of the OECD. The opinions expressed and arguments employed herein do not necessarily reflect the official views of OECD member countries.

This document and any map included herein are without prejudice to the status of or sovereignty over any territory, to the delimitation of international frontiers and boundaries and to the name of any territory, city or area.

Please cite this publication as:
OECD (2015), *Drying Wells, Rising Stakes: Towards Sustainable Agricultural Groundwater Use*, OECD Studies on Water, OECD Publishing, Paris.
http://dx.doi.org/10.1787/9789264238701-en

ISBN 978-92-64-23869-5 (print)
ISBN 978-92-64-23870-1 (PDF)

Series: OECD Studies on Water
ISSN 2224-5073 (print)
ISSN 2224-5081 (online)

Co-published by IWA Publishing
Alliance House, 12 Caxton Street, London SW1H 0QS, UK
Tel: +44 (0)20 7654 5500, Fax: +44 (0)20 7654 5555
publications@iwap.co.uk
www.iwapublishing.com

ISBN: 9781780407968 (Paperback)
ISBN: 9781780407975 (eBook)

The statistical data for Israel are supplied by and under the responsibility of the relevant Israeli authorities. The use of such data by the OECD is without prejudice to the status of the Golan Heights, East Jerusalem and Israeli settlements in the West Bank under the terms of international law.

Photo credits: Cover © Eddie J. Rodriguez/Shutterstock.com; © Taro Yamada/Corbis

Corrigenda to OECD publications may be found on line at: *www.oecd.org/about/publishing/corrigenda.htm*.

Foreword

As a natural reserve relatively resilient to climate variability, groundwater has provided large benefits to irrigated agriculture in semi-arid OECD countries. It has supported the development and expanded production of commodity crops in the US Midwest and Mexico and high value products in semi-arid areas of Mediterranean Europe or the Middle East. But intensive use beyond recharge capacity in certain regions has depleted resources and increasingly generates significant negative environmental externalities, including stream depletion, saline intrusion and land subsidence.

The report studies the challenges of managing groundwater use in agriculture sustainably, acknowledging its increasing importance as a tool for agriculture's adaptation to climate change. It provides new data on the status of groundwater irrigation, proposes a characterisation of groundwater agricultural systems, assesses the economic effects of existing management instruments and analyses the range of policies used in OECD countries.

The study builds on OECD's work on water, especially the 2010 report overseeing issues around the sustainable management of water resources in agriculture, the 2014 report on climate change, water and agriculture, and the 2015 survey-based analysis of water resource allocation regimes in OECD countries.

The analysis relies on new information collected through a comprehensive questionnaire of groundwater management policies in OECD countries and selected regions launched in the summer of 2014. Contributions from OECD delegations and OECD country experts in responding to this questionnaire are gratefully acknowledged.

This report was written by Guillaume Gruère. Chapter 3 is based on a consultant report written by Nicholas Brozović, Robert B. Daugherty Water for Food Institute and University of Nebraska-Lincoln, United States. Noura Takrouri-Jolly provided statistical support at different stages of the project and contributed significantly to the design, implementation and analysis of the OECD questionnaire. The report also benefited from comments and suggestions from Dale Andrew, Nicholas Brozović, Carmel Cahill, Anthony Cox, Kathleen Dominique, Jane Ellis, Julien Hardelin, Franck Jésus, Hannah Leckie, Xavier Leflaive and Janine Treves. Michèle Patterson edited and formatted the report. Françoise Bénicourt and Stéphanie Lincourt helped manage the administrative process.

Table of contents

Tables

Figures

Boxes

Executive summary

Groundwater resources sustain a significant and increasing share of irrigated agricultural production. On a global scale, groundwater represents over 40% of consumptive irrigation water use, covering just under 40% of irrigated land globally. In OECD countries, groundwater for agriculture irrigation is used on 23 million hectares, for an estimated annual volume of 123.5 km^3, accounting for about 20% of global irrigation withdrawals.

These aggregate figures, however, mask a large heterogeneity across OECD countries. Groundwater is primarily used for irrigation in semi-arid areas in around ten countries located mostly in North America and the Mediterranean region. At the same time, some other OECD countries, including those located in more humid agro-climatic zones, do not use groundwater significantly for agriculture.

The significant short-term advantages groundwater use confers for irrigators, coupled with the growing demand from other sectors, have contributed to increase its use beyond natural recharge in some OECD regions. Agricultural groundwater irrigation expansion can largely be explained by its relative insulation from climate variation and the ability to provide water on demand to individual farmers that can access it. Yet, these advantages have contributed to the intensive use of groundwater resources beyond recharge in several semi-arid regions, lowering water tables with short-term and long-term consequences for farmers.

Intensive groundwater pumping for irrigation has generated large negative externalities affecting agriculture, other users and the environment. Intensive groundwater use in some areas of OECD countries has resulted in stream depletion, with repercussions on surface users and related ecosystems. The salinization of coastal aquifers, sometimes irreversible, has affected crop choice for agriculture and ecosystems. Agricultural intensive withdrawals of groundwater are also responsible for land subsidence (sinking) in some regions, damaging infrastructures in urban and rural areas at a very high cost.

With climate change expected to induce increased water stress in more OECD regions, groundwater issues will become more pressing. Surface water volatility and weather shocks will greatly expand the role of groundwater in current and future potentially irrigated areas. As a result, several regions in OECD countries that do not currently use agriculture groundwater significantly will likely do so in the future at the risk of facing the same challenges being experienced in regions which already use groundwater intensively.

Groundwater management has a role in redressing these externalities, and transforming groundwater from being a productive input for agriculture to a long-term, climate insulated, sustainable reservoir, wherever possible. If well managed, groundwater can and should act as a powerful climate adaptation option, a natural insurance mechanism, and not just a component of freshwater supplies.

This requires first an understanding of the heterogeneity of agricultural groundwater systems; a generic characterisation is proposed to differentiate and analyse management and policy responses in OECD countries. Groundwater is essentially a local resource, whose characteristics greatly depens on specific conditions and use at the aquifer level. Four factors can help characterise agriculture groundwater systems: agro-climatic conditions, access to and availability of surface water, availability of accessible and usable groundwater resources, and trends in use and profitability of groundwater irrigation relative to other uses.

Second, policy instruments need to be selected to respond to the defined characteristics and challenges. The economic literature shows that no single policy instrument can address groundwater management challenges in all settings; each type of instrument has advantages and drawbacks.

There is a wide diversity of policies applied to manage groundwater use in agriculture in OECD countries, often only partially correlated with specific regional constraints. Policies are founded on different legal systems; they focus on the demand side, supply side or both, and use direct or indirect approaches to regulatory, economic or collective management. While there is no visible link between the scope of management and the intensity of constraints, economic and supply-side approaches are more prevalent in areas under higher agricultural groundwater stress.

How should these policies evolve to help improve agriculture groundwater management?

- Six general conditions are identified for successful management: a) build and maintain sufficient knowledge of groundwater resource and use; b) manage surface and groundwater conjunctively (together) where relevant; c) favour instruments that directly target groundwater use over indirect measures (e.g. land use regulation), where possible; d) prioritise demand–side approaches, e) enhance the enforcement of regulatory measures (e.g. water entitlements) before moving to other approaches; and f) avoid non-water related price distorting policy measures, such as subsidies towards water intensive crops and energy, that could affect groundwater use.
- Policies should be constructed as a "tripod", combining regulatory, economic and collective management instruments. Groundwater entitlements systems should remain the core of groundwater management. Collective action- based approaches are present in many of the successful cases to redress externalities. Economic instruments can support efficient solutions to groundwater scarcity and depletion problems.
- Measures that increase agricultural water productivity and support new recharge mechanisms, such as aquifer storage and recovery, *provide complementary tools* in cases of high water stress.

This three-part package should be adapted to locally-specific agriculture groundwater systems, which may call for the division of management into functional subunits.

Survey results indicate that these recommendations have not been uniformly applied in OECD countries or regions that use groundwater intensively for agriculture. In particular, there seems to be a relatively low level of knowledge on groundwater resources and use. Most OECD countries or regions in the survey sample have also applied incomplete management schemes, missing part of the "tripod" recommended approaches.

- Improving information systems on groundwater resources and flows should be the priority for all countries using or planning to use groundwater for irrigation. Long-term groundwater depletion and externalities cannot be managed without information on groundwater resources and use. Lack of information also makes efficient instruments more difficult to design and to implement.
- Configuring and enforcing a balanced set of locally adapted management instruments will be necessary to ensure that groundwater can play its role to support sustainable agricultural production. Incomplete policies that are poorly enforced, or are relatively rigid in their implementation, may prevent the sustainable exploitation of groundwater for agriculture in the future.

Chapter 1

The worrisome trends in groundwater irrigation expansion

This chapter presents the challenges of groundwater management, and provides an overview of the status and use of groundwater for agriculture in OECD countries. It examines recent data, trends, and indicators of groundwater use and stress at the national, regional and aquifer levels, and reviews the evidence on the expected effects of climate change.

The statistical data for Israel are supplied by and under the responsibility of the relevant Israeli authorities. The use of such data by the OECD is without prejudice to the status of the Golan Heights, East Jerusalem and Israeli settlements in the West Bank under the terms of international law.

Key messages

The spectacular expansion in groundwater irrigation in the past four decades, called a "silent revolution", has made groundwater indispensable for agriculture production in many countries. This expansion can largely be explained by the capacity of groundwater to act as a reliable water source for agriculture irrigation, providing water "on demand", while being largely unaffected by surface hydrological variation in the short term.

This expansion, however, has led to the use of groundwater beyond natural recharge in many regions, in some cases with significant negative economic and environmental impact. Continued abstraction results in the lowering of water tables, which in turn increases the cost of pumping and could possibly create a "race to the bottom" among producers. Environmental effects with direct consequences on agriculture production are also produced, and will affect the future use of this resource for agriculture.

Within the OECD area, groundwater irrigation was used on an estimated 23 million hectares in 2010, which represents one-third of the OECD total irrigated area. As of 2010, an estimated annual volume of 123.5 km^3 was used for irrigation in OECD countries, accounting for 56% of total groundwater withdrawals in the same countries, or about 20% of total groundwater irrigation globally. At the same time, both irrigated areas and groundwater withdrawals for agriculture were found to vary widely, from almost no use to intensive use of groundwater in agriculture.

In some of these countries, particularly in parts of North America and the Mediterranean region, agriculture use of groundwater is growing and contributing to groundwater stress. The leading groundwater irrigating countries in OECD have increased their use of groundwater over the past 25 years, while others have a relatively stable groundwater use. The OECD average groundwater development stress (GDS) for agriculture at the national level, which measures the ratio of groundwater extraction over natural recharge, was estimated at 7.6% in 2010, with large variations across OECD countries, ranging from zero to over 100%. More variation is observed at the regional and aquifer levels.

With climate change expected to induce increased water stress in more OECD regions, groundwater issues will become more pressing. Surface water volatility, reduced aquifer recharge, and weather shocks will greatly expand the role of groundwater in current and future potentially irrigated areas. At the same time, climate change is expected to induce higher salinity and reduce aquifer recharge, including in some of the major aquifers used by agriculture in OECD countries, creating further pressure.

The large variations in use and evolving constraints across and within OECD countries call for a better understanding of agricultural groundwater systems before addressing potential policy responses. A characterisation of groundwater systems is proposed in Chapter 2.

The increasing significance and challenges of groundwater irrigation

In a context of increased demand for food, competing water demands, and growing climate variability, the management of water is expected to play an increasingly critical role in agriculture (OECD, 2010b and 2014a). Calls for agriculture production growth to respond to population increase will increase demand for water, which will be even greater due to expected high demand from other sectors. The provision of freshwater resources will also continue to be directly affected by climate change, and agriculture's adaptation to new climatic conditions will depend highly on water management (OECD, 2014a and 2014b). These three growing tensions, increased demand for food, climate change, and competition for access, operating in parallel, call for the efficient and sustainable use of all water resources in agriculture; surface water and the perhaps less mentioned but much larger body represented by groundwater.

Groundwater is constituted of water that fully saturates all fissures and pores, and is contained in an aquifer matrix located beneath the surface, as opposed to free surface water bodies like streams, reservoirs, or lakes (e.g. Giordano, 2009; Siebert et al., 2010. A full glossary is available at the end of the report).[1] As such, it occupies a specific part of the water cycle, connected to, but often semi-independent from, surface water. While it is often used as a complement to surface water, groundwater is a significant source of public water supply; about 60% of total drinking water is used for human consumption (Margat, 2008). It also provides crucial support to agriculture and industrial activities in multiple countries. Overall, over 2.6 billion people may rely on groundwater resources (OECD, 2013a).

Groundwater irrigation expansion: A silent revolution

Groundwater represents a major share of water used for agriculture irrigation (Giordano and Villholth, 2007). More than 60% of groundwater use is for agriculture in semi-arid and arid regions which produce 40% of the world's food (Morris et al., 2003; OECD, 2013a). Globally, Shah et al. (2007) estimated that agriculture groundwater supported an annual output equivalent to USD 210-230 billion, corresponding to an average gross productivity of USD 0.23- 0.26/m^3 abstracted. Total consumptive groundwater use for irrigation in 2010 was estimated at 545 km^3/year, or 43% of the total consumptive irrigation water use[2] (Siebert et al., 2010). The total area used for groundwater irrigation covered 98 million ha, or 39% of the total irrigated land in 2010 (Siebert et al., 2010). Given that these estimates focus on groundwater sources alone, they are likely undervalued because groundwater is very often used in conjunction with surface water for irrigation (Kemper, 2007).

Groundwater resources have allowed major gains in global agriculture productivity and continue to sustain a significant share of global crop production (OECD, 2012a and 2012b). The spectacular expansion in groundwater irrigation in the past four decades has been termed a "silent revolution", resulting in large effects on agriculture production levels (Garrido et al., 2006). Groundwater development for irrigation developed significantly first in Italy, Mexico, Spain and the United States and was followed by rapid expansion primarily in Asia (Shah et al., 2007; van der Gun, 2012). It now accounts for half of South Asia's irrigation and supports two-thirds of grain crops supplied in the People's Republic of China (Giordano and Villholth, 2007). It also plays a significant role in agriculture in OECD countries, especially those with arid or semi-arid conditions. Over 60% of irrigated agriculture and nearly half of the farmers use groundwater in the United States (Gollehon and Quinby, 2006; Scanlon et al., 2012). It represents over 70% of Spain's irrigation, providing five times more value and three times more jobs than irrigation from surface water in the region of Andalucía (Hernandes-Mora et al., 2003). Groundwater also provides a third of the water used for irrigation in Mexico, which is the largest user of groundwater in Latin

America, with over 100 000 large-capacity pumps (Scott et al., 2010). Its use in agriculture in Australia is estimated to contribute AUD 11 billion annually to the economy (Deloitte Access Economics, 2013).

The scope of use and significance of groundwater can largely be explained by its intrinsic physical characteristics.[3] First, unlike surface water, it is characterised by high storage capacity relative to inflows (Giordano, 2009). Second, it flows at a much slower pace than surface water (OECD, 2013e). Third, its quality is generally superior to that of surface water, in particular with regards to bacterial contamination (hence its importance for drinking water). Due to its relative insulation from weather changes, it constitutes a type of "buffer storage" that can complement surface water (Morris et al., 2003). Indeed, the low rates of inflows and outflows to groundwater reserves assure the viability of the resource even in times of drought (Bovolo et al., 2009; OECD, 2013e).

Groundwater is effectively used in agriculture as a natural storage facility, acting as insurance against drought (Garrido and Iglesias, 2006), or a "water savings account" enabling producers to sustain the use of water when surface water is not sufficient. For instance, Howitt et al. (2014) estimated that the 2014 drought in California would result in a loss of 6.6 million acre-foot in surface water, of which 5 million acre-foot could be recovered via groundwater pumping. In arid and semi-arid areas, groundwater irrigation also provides longer growing seasons and lower risks of pest and disease (Siebert et al., 2010). Its capacity to serve as a reservoir also makes it an important tool to increase long-term resilience to climate change (Green et al., 2011; Gleeson and Cardiff, 2013; OECD, 2013d).

Groundwater is also used, even in less climate-stressed areas, by individual farmers of small to large size farms due to its ability to provide "water on demand", i.e. letting producers manage their water depending on their needs (OECD, 2010b). Groundwater resources are characterised by their *horizontal* dimension, with aquifers covering large areas not always contiguous with surface water basins, allowing farmers to access water under their field in a rather equitable manner (Kemper, 2007). In multiple areas, with shallow aquifers, it is furthermore easy and relatively cheap to access thanks to the development of affordable pumping technologies. As a result, it is seen by farmers as an attractive, reliable, and easily accessible source of water (Garrido and Iglesias, 2006) and is highly popular for this reason (Garduño and Foster, 2010). Indeed, several studies in different countries have shown the systematic preference of farmers for groundwater irrigation (Shah, 2008).

Leading to increasing pressures in areas of intensive irrigation

These advantages, however, have contributed to increase its use beyond natural recharge in many regions. While agriculture is a significant contributor to the recharge of shallow aquifers, both via surface and groundwater irrigation (Taylor et al., 2012), it has increasingly become an even larger withdrawer of groundwater. Non-renewable abstraction reached 234 km^3/year or 20% of gross irrigation demand in 2000, and had more than tripled since 1960 (Wada et al., 2012). Due in part to increased climate variability, which affects access to surface water, groundwater resources are increasingly used to the point of being exploited beyond recharge in multiple agricultural regions (Taylor et al., 2012).

Such intensive use of groundwater resources, otherwise known as "overdrafting",[4] will affect users of groundwater, including irrigated agriculture. This situation has made groundwater use "*one of the most important challenges for agriculture*" (OECD, 1998; 2013e). As noted by FAO (2011), "*because of the dependence of many key food production areas on groundwater, declining aquifer levels and continued abstraction of non-renewable groundwater present a growing risk to local and global food production*." Continued abstraction results in falling water tables, which in turn increases the cost of pumping and can create a "race to the bottom" among producers. In Mediterranean countries, aquifers that contribute largely to drinking water supplies are used by farmers beyond their recharge rates with consequences for both types of uses (OECD, 2011b). Overdrafting will also likely affect countries where groundwater is not a major source of irrigation via market linkages; the gradual depletion of groundwater resources especially in South and East Asia on agricultural land that feeds hundreds of millions of people may have global food security consequences with trade and production implications (Wada et al., 2012).

Groundwater overdrafting in rural areas also affects the environment (OECD 1998; 2011a) and can generate environmental effects with direct consequences on agriculture production. In particular, it can affect rivers, lakes and other streams, and generate land subsidence and increased salinity (Bovolo et al., 2009). Groundwater pumping can result in desiccation of natural reserves, as seen in the Netherlands, and contribute to the drying up of wetlands, as seen in southern Europe, with significant loss for water quality filtering (Hellegers et al., 2001; UNECE, 2011). In Mexico and western United States, it has resulted in significant land collapses (Foster, 2008; Sneed et al., 2013). Intensive pumping in coastal aquifers, or aquifers connected to saline water bodies, is a significant source of salinization of groundwater, affecting the crop choice for agriculture and ecosystems, in particular in the wetlands, rivers, ponds, springs and streams to which it is connected (Schoengold and Zilberman, 2007; Fuentes, 2011; UNECE, 2011; Amores et al., 2013).[5]

Lastly, overdraft of groundwater resources can affect the future use of this resource for agriculture. Farmers using groundwater beyond recharge may lose a source of future income (an option value, e.g. Howitt et al., 2014). It is estimated that 97% of the groundwater, via run-offs, evapotranspiration and subsequent precipitation, ends up in oceans (Wada et al., 2010) and a certain lapse of time, varying from a few years to several millennia, can be needed to recharge the aquifer, assuming its capacity remains unchanged. In extreme cases, especially under arid conditions where surface water is not easily available, situations of "boom and bust" could result in agriculture, reaching a no return water level, under which it is no longer profitable to farm.

Identifying policy solutions to address the growing and diverse challenges of groundwater resource management in agriculture in OECD countries

A number of OECD reports have looked at specific aspects of groundwater policies as part of more general studies of water resource management. However, none of these reports have specifically addressed the intrinsic challenges such policies are faced with. Reports on water have, for instance, provided general principles for sustainable water management (OECD, 2010b) and discussed the use of economic instruments (OECD, 2013e). Groundwater has been discussed in the context of pricing and financing (OECD, 2009a and 2009b), energy (OECD, 2012b), risk management (OECD, 2013e), and broader perspectives covering climate change (OECD, 2013d and 2014a). Groundwater is also featured in the reviews of water reforms at the country level (e.g. Fuentes, 2011; OECD, 2013b). All these reports include sections, sub-sections, paragraphs, or illustrations that relate to groundwater, but they do not convey policy conclusions specifically geared towards the managers of specific types of groundwater, especially in the context of agriculture.

Several common threads can be identified in the broader literature on agriculture and groundwater. First, a consistent observation is that groundwater is generally under-studied and there is a need for more in-depth assessment of groundwater stocks, use, and management practices. Insufficient knowledge on resource flows and management practices is problematic in addressing pending challenges in a number of regions (e.g. Struzik, 2013). Second, groundwater policies are identified as requiring further in-depth analysis (Koundouri, 2004; OECD 2010b). Finally, specific types of aquifer and their associated constraints are emphasised as having a critical role in the determination of sustainable management plans for agriculture (Giordiano and Villholth, 2007).

This report will look at quantitative groundwater management challenges for agriculture and the potential role of policies, where possible, while accounting for the different types of groundwater systems. The objective is to provide a comprehensive analysis of the economics and policies to find responses to the growing and diverse challenges of groundwater resource management in agriculture in OECD countries by focusing in particular on long-term groundwater depletion and associated externalities.

There are several caveats. First, the analysis will focus largely on groundwater use associated with irrigated agriculture. This means in particular that water consumption related to livestock production will not be extensively analysed, notably because of the lack of data and information. Second, the management of excessive groundwater levels (and groundwater flooding) not related to agriculture use but affecting agriculture will be left out of the discussion.[6] Third, while the report looks at externality induced from groundwater use in agriculture, it does not explicitly focuses on the preservation of environmental flows needed for the ecosystems to thrive. Fourth, challenges associated with groundwater quality will not be discussed either, except in the case of salinity induced from groundwater pumping. There are clear links between quality and volume, notably via concentration, but those will be left out of this report.[7] This caveat does not imply that quality issues are not important; in fact some may be emerging, and their importance is acknowledged especially where groundwater is used for drinking purposes (OECD, 2012c). Groundwater-dependent ecosystems may also be affected by any change in water quality. A more in-depth analysis of agriculture-induced quality issues, including nitrogen filtration into aquifers and the quality-quantity interaction, is left for future endeavours. Fifth, as noted above, not all OECD countries are concerned by the challenges outlined in this report, but some may be in the future. Lastly, the policy recommendations identified in this report may not be applicable in many countries on their own at the national level; they will need to be part of broader water allocation reforms as defined and analysed in OECD (2015b). Therefore, they do not aim to supplant broader reform efforts, but to complement them by providing specific management instruments for groundwater especially in regions with intensive use.

This report is organised in five chapters. Chapter 1 provides a synthetic review on where OECD countries stand in terms of groundwater resources and agricultural use, stress indicators and projections under climate change. Chapter 2 reviews the specific characteristics of aquifers, and the main issues they currently face and will face in the future. Chapter 3 reviews groundwater management policies based on a 2014 OECD country survey. Chapter 4 reports the policies in place in OECD countries, and the final chapter identifies gaps and proposes potential improvements for agriculture groundwater policies.

Groundwater use in agriculture accounts for over half of OECD countries' total groundwater withdrawals, with large differences across countries

The challenges of collecting information on an "invisible" resource

Groundwater is often considered an "invisible" resource due to its largely non-visible nature, the complexity of its hydrogeological processes and specialised knowledge requirements, the intrinsic specificity of each aquifer, and the difficulty of measuring its state and flows (Monginoul and Rinaudo, 2013).[8] Various approaches, even if imperfect, have been developed and used to assess groundwater quantity and flows. Yet the lack of investment in measurement and of overall groundwater expertise in water resource management groups — which tend to be more focused on surface water issues — make the assessment of groundwater resources often incomplete in several OECD countries (Struzik, 2013). At the regional level in Mediterranean countries, there is a critical lack of data on total groundwater use, number of boreholes, groundwater quality changes (including salinity), groundwater costs and prices, and interaction with surface water (EASAC, 2010a).[9] The fact that measuring groundwater resource availability is difficult and expensive, and that the measurement of groundwater use is either not applied or that data are not always shared in many regions, sometimes for political reasons, contributes to the general apparent lack of reliable knowledge on its status and uses (BGS, 2009).

Even if groundwater flows are easier to measure than stocks given the intrinsic complexity of aquifer structures, it is very challenging to assess precisely the flows in and out of aquifers in a comprehensive manner (Giordano, 2009). On the discharge side, natural uses and flows operating both vertically and horizontally in aquifers increase the complexity of the picture. The lack of monitoring and/or reporting on pumping in many countries also plays a significant role, especially in agriculture. Different types of monitoring tools can lead to divergent results, as observed in the case of irrigation in Arizona (Cohen et al., 2013). At the same time, recharge measurement is very difficult given the differences in situation, soil profiles and soil covers, and connections with surface water bodies.[10] In many cases, field crop activities are known to actively participate in the recharging of groundwater, even sometimes more significantly than natural ecosystems (Taylor et al., 2012), but specific local constraints once again make generalisations challenging.

Measurement efforts that combine different types of local and regional measurements, including the use of satellite-based tools, have improved assessments, but the overall picture remains imperfect. Satellite data, including that generated by the US NASA's Gravity Recovery and Climate Experiment (GRACE), have been able to provide overall monthly and yearly changes in groundwater stock in multiple regions. It demonstrated, for example, the diminution of resources in California over time (e.g. Famiglietti et al., 2011), but given that the resolution satellite-generated data remains insufficient, it could only provide assessments over significant periods of time and its use requires complementary traditional measurements.

The picture on available resources and data is therefore mixed, particularly at the national level. Smaller countries, for which groundwater is a major source of freshwater and which have more homogeneous geological profiles of agricultural land, and where data is collected and shared, are able to properly track groundwater resources over time. An example is Denmark's National Groundwater Mapping and Management program, financed by private and public water consumers, which uses new tools to track groundwater (DWF, 2012). However, larger countries may have more general information overall given the more detailed monitoring of groundwater hotspots (e.g. United States). Lastly, precipitation-abundant countries for which groundwater is not as important a resource, and which have predominantly rain fed agriculture, do not dedicate many resources to groundwater quantity measurement, resulting in a general lack of information (e.g. see Council of Canadian Academies, 2013:93).

With these caveats, the following section provides information on groundwater resources and agricultural use based on data collected via an OECD country questionnaire launched in the summer of 2014 (see Chapter 3 for details), complemented by available secondary data which relies mostly on estimates and assumptions rather than actual measurements, [11] as well as consultations with several water experts. [12]

Agriculture groundwater use in OECD countries: From non-users to major irrigators

Table 1.1 provides ranges of estimates for global groundwater stocks, inflows and withdrawals based on different analytical estimates of water resources (Margat and van der Gun, 2013). Overall, groundwater represents a major portion of usable water resources, accounting for 96% of liquid freshwater (UNESCO, 2008). Groundwater reserves are estimated to be over 20 million km^3, of which 40% is freshwater. Annual recharge amounts to around 12 000 km^3, while annual withdrawal ranges between 600-1100km^3, with recent estimates ranging between 950 and 1000 km^3 (OECD, 2009b; van der Gun, 2012; Margat and van der Gun 2013).[13]

Agriculture uses groundwater primarily for irrigation, but also as a drinking resource for livestock and agricultural product processing. Most available figures on groundwater use, however, focus solely on irrigation; hence the emphasis on irrigation in this analysis (see Annex 1.A for livestock data, which is often limited, as noted in Deloitte Access, 2013). Globally, irrigation accounts for two-thirds of groundwater withdrawals, with total estimates ranging between 545 and 688 km^3/year (Siebert et al., 2010; Margat and van der Gun, 2013). Taking the 545 km^3/year estimate, this accounts for about 43% of total irrigation.

Despite evidence of local water risks in the future, these estimates show that the global use of agricultural irrigation is relatively insignificant, representing about 5% of annual natural inflows, and only 0.003% of total groundwater reserves. At the same time, Table 1.1 also shows that agriculture accounts for a very large share of total withdrawals, amounting to an estimated 70% in 2010. This aggregate estimate, however, combines very different rates across countries and equally variable rates at subnational levels (Margat and van der Gun, 2013).

Competing uses include industry and public water supplies; in many regions, groundwater is the sole source for drinking water. As shown in Figure 1.1, agriculture use represents a significant portion of groundwater use (more than 45%) in at least nine OECD countries: Israel, Chile, Korea, Turkey, Australia, the United Sates, Spain, Mexico and Portugal. Domestic use (drinking water) represents the other main user in these OECD countries, and is especially prominent in European countries (Finland, Slovenia, Estonia, Austria, Denmark and the United Kingdom). In contrast, groundwater use in those countries is relatively limited in the industry and energy sectors.

Estimates of groundwater irrigated areas and shares of total irrigated areas in OECD countries are shown in Figure 1.2 for the year 2010 (and in Annex 1.B, Table 1.6). Groundwater irrigated area in OECD countries covers about 23 million hectares, or about 26% of the global irrigation figures (Siebert et al., 2010). More than half of OECD's groundwater irrigation is located in the United States. Mexico, Turkey, Spain, Italy, France, Greece and Australia follow, with at least 500 000 ha each of groundwater irrigated fields.

Table 1.1. Estimated global groundwater stocks, inflows and withdrawals

	Type	Estimates (km^3)	Share of total stock
Stocks	Freshwater	8 to 10 million	~40%
	Brackish or saline water	12 to 14 million	~60%
	Total	20 to 24 million	100%
Annual recharge	Total	11 to 15 thousand	0.05-0.08%
Annual withdrawal	Total	0.6 to 1.1 thousand	0.0025-0.0055%
	Agriculture irrigation only	0.545-0.688 thousand	0.0023-0.0034%

Source: Margat and van der Gun (2013), based on a review of existing estimates for stocks and recharge, Siebert et al. (2010), van den Gut (2012) and Margat and van der Gun (2013) for agriculture.

Figure 1.1. Proportion of groundwater use by sector in OECD countries (2008-2013)

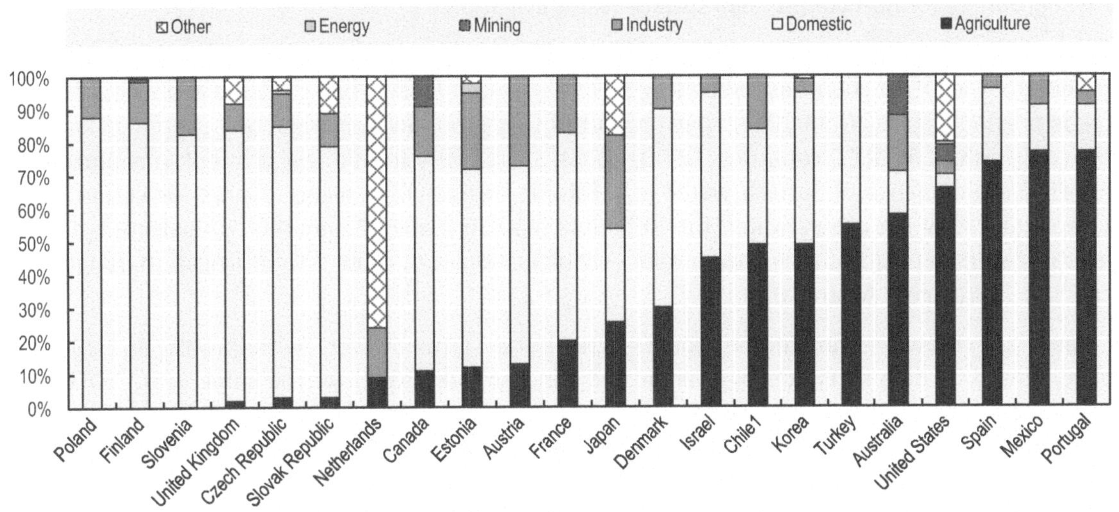

Note: The agriculture sector *a priori* includes livestock.1.Data refer to the year 2003.
Source: 2014 OECD questionnaire on groundwater use in agriculture.

Figure 1.2. Area irrigated with groundwater and proportion of total irrigated areas in OECD countries (2010)

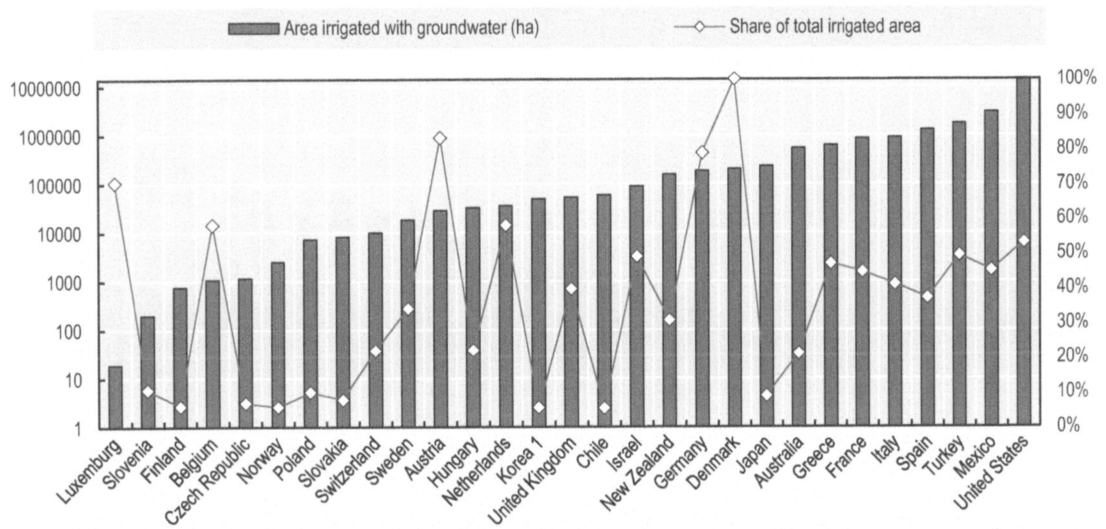

Notes: The figure uses a logarithmic scale for visibility. Canada and Portugal are excluded for lack of available comparable data.

1. Korea's groundwater irrigated area includes only paddy rice and may be underestimated.

Source: IGRAC (2012), ggmn.e-id.nl/ggmn/GlobalOverview.html; 2014 OECD Questionnaire on groundwater use in agriculture.

Groundwater irrigation is used to support multiple agriculture activities. In the United States, irrigated crops that use groundwater significantly include cereals (rice, wheat, maize), oilseeds (soybean and cotton), and specialty crops (vegetables, fruit, nuts), for which groundwater serves as a secondary water source especially in drought years (NGWA, 2013; Esnault et al., 2014; see Annex 1.A). It is also used for sugarcane, cotton, rice and nut trees in Australia (Deloitte Access, 2013), and olive trees, vineyards, and for greenhouse fruits and vegetables in Spain and Greece (Garrido et al., 2006; Molinero et al., 2011; EASAC, 2010b). The sanitary and technical requirements associated with certain irrigation techniques — including water-drip systems, which necessitate avoiding sedimentary residues for risk of blocking — makes groundwater relatively more appropriate for horticulture than surface water irrigation.

Based on these 2010 estimates, groundwater irrigated land covers on average 33% of the total irrigated areas in OECD countries, exceeding 30% of irrigated land in half of OECD countries. This exceeds previous figures from 2002, which reported that groundwater irrigation covered at least 30% in only a third of OECD countries (OECD, 2006). Shares vary from a few per cent in countries with relative abundance of surface water (e.g. Estonia, Norway), to higher shares for countries relying significantly on groundwater for irrigation (e.g. Germany, Denmark). As expected, the largest irrigators in terms of area have a relatively higher share of groundwater irrigation, ranging from 35 to 60% (except Australia).

Figure 1.3 shows estimated national agricultural groundwater withdrawals in 2010 complemented by data estimates by Margat and van der Gun (2013) for the same year.[14] OECD countries' aggregate groundwater withdrawals as of 2010 are estimated to be 221.5 km^3/year, of which 56% or 123.5 km^3/year was used for irrigation purposes. These irrigation withdrawals represented between 18% and 23% of the global groundwater irrigation withdrawals at the global level for the same year. The United States, Mexico, Turkey, Italy, Spain and Greece lead in irrigation withdrawal volumes. Countries from northern and central Europe, including Poland, Switzerland, Norway, Sweden and Estonia, use virtually no groundwater for agricultural irrigation.

Figure 1.3. Estimated groundwater abstraction for agriculture irrigation (2010)

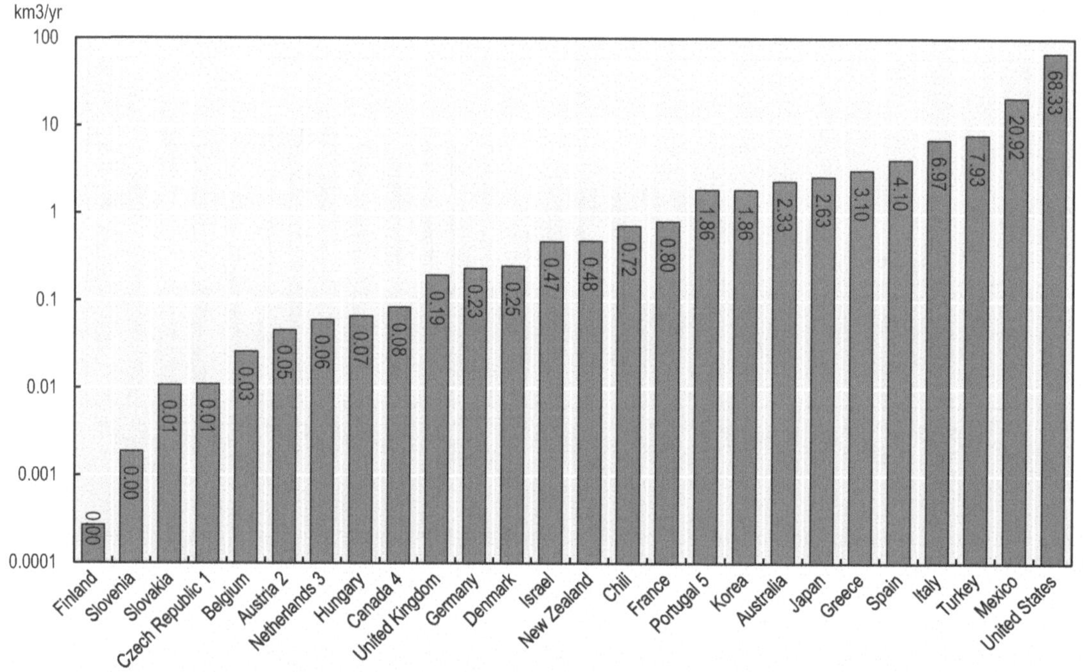

Notes: This figure uses a logarithmic scale for visibility. These abstraction figures are presented in gross terms- they do not account of recharge associated with agricultural activities. Estonia, Iceland, Luxemburg, Poland, Norway, Sweden and Switzerland are not included as they report to have zero or negligible irrigation.

1. The agricultural share for Czech Republic is based on OECD (2013c).
2. Austrian data is based on 2008-10 data.
3. Data refer to the year 2011.
4. Data refer to the year 2012.
5. Data refer to the year 2009.

Source: 2014 OECD questionnaire on groundwater use in agriculture and Margat and van der Gun (2013).

To provide a more consistent cross-country comparison, Figure 1.4 presents the ratios of estimated groundwater use over groundwater irrigated areas for selected OECD countries to help distinguish the relative intensity of groundwater use in irrigation. Six countries stand out with ratios significantly over the OECD average. Of these, Mexico, Italy, and Japan are also relatively large users. Portugal also presents important irrigation intensity, but without specific information for "groundwater irrigation area", not being the reference values (national statistics) comparable to the present study.[15] The United States, Turkey, Greece, Spain and Australia, on the other hand, despite being among the largest users, have a groundwater irrigation intensity under the estimated OECD average of 5 194 m^3/ha in 2010.

Figure 1.4. Groundwater abstraction for agricultural use, by area, in OECD countries (2010)(m^3/ha)

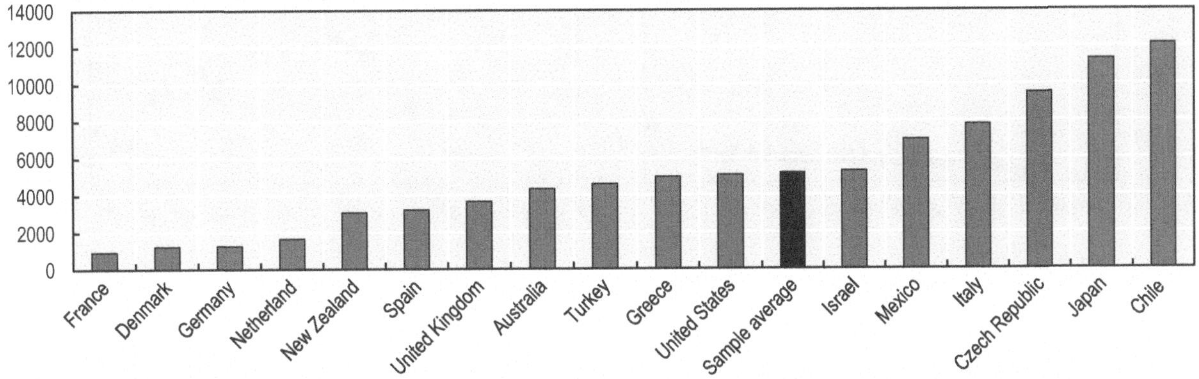

Note: Korea (due to underestimated area), Portugal (due to unavailable total area) and countries using less than 0.1 km^3/year are not included.

Source: Derived from 2014 OECD questionnaire, Margat and van der Gun (2013) and IGRAC data for 2010 (ggmn.e-id.nl/ggmn/GlobalOverview.html.). See Annex 1.B for details.

Historical trends may also provide complementary information. Figure 1.5 shows the trends in use in selected OECD countries with a relatively high level of agricultural groundwater use between 1985 and 2010. It provides a good representation of the incompleteness of data series. The United States, Mexico and Turkey are separated on the left panel for scale purposes. These three countries have increased their use over the 25-year period, although at a fluctuating pace in the United States. Spain and Greece (right panel) have also increased their use significantly, if at a less steady pace. Japan has significantly reduced its agricultural groundwater use, and Australia's use appears relatively stable in recent years. Part of these changes may be due to variations in climatic conditions and surface water availability and access on the one hand, and trends in competing uses on the other.[16]

In some countries, specific assessments also exist at the aquifer level, with .much of the interest focused in particular on the 37 "major" aquifers, defined as large and voluminous aquifers. Box 1.1 provides an overview of the main types of aquifers and their implication on groundwater flows; not all aquifers are considered renewable, but this qualification has only a relative meaning. Table 1.2 provides the basic characteristics of the eight major aquifer systems present in OECD countries.

Figure 1.5. Trends in total groundwater withdrawal for agriculture in selected OECD countries (1985 -2010) (km3/year)

The figures include livestock and other agricultural uses where available

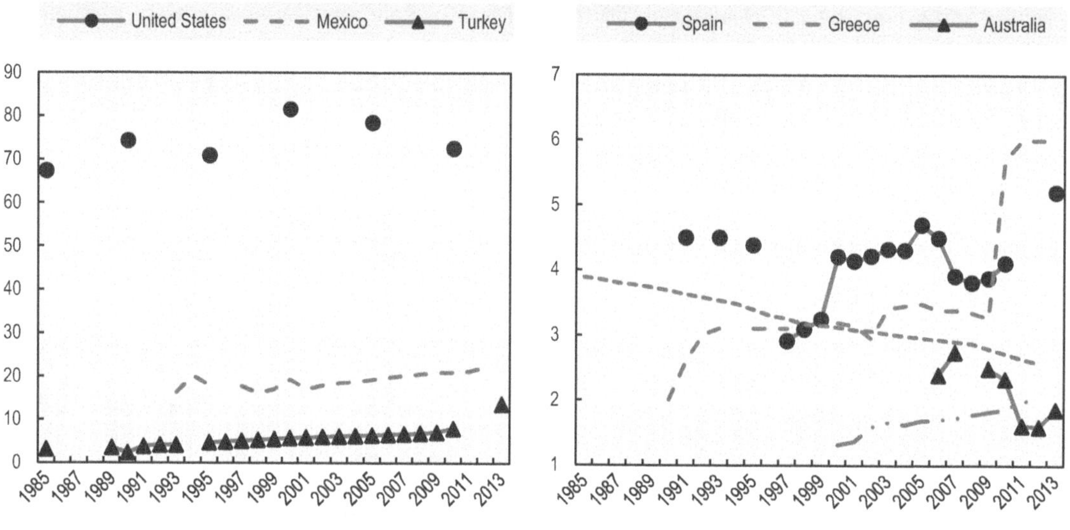

Source: OECD (2013c), 2014 OECD questionnaire on groundwater

and Margat and van der Gun (2013) for 2010.

Table 1.2. Major aquifer systems in OECD countries

Country or countries	Name	Area ('000 km^2)	Maximum thickness (m)	Theoretical reserves (km^3)	Recharge rate (km^3/year)
Australia	Great Artesian Basin	1 700	3 000	65 000	1.1
	Canning Basin	430	1 000	n.a.	n.a.
France	Paris Basin	190	3 200	500-1000	20-30
Canada and United States	Northern Great Plains Aquifer	~2 000	n.a.	n.a.	n.a.
Mexico and United States	Atlantic and Gulf Coastal Plains Aquifer	1 500	12 000	n.a.	n.a.
United States	Cambrian–Ordovician Aquifer System	250		n.a.	n.a.
	Californian Central Valley Aquifer System	80	600	1130	7
	High Plains (Ogallala Aquifer)	450	150	~15 000	6-8

Note: n.a. Not available.
Source: Margat and van der Gun (2013); www.environment.gov.au/water/publications/watermatters/water-matters-may-2009.html.

Box 1.1. Basic elements of groundwater flows

Groundwater is contained in the saturated water layer of the earth beneath aquifers' ceiling (the surface of the saturated layer). The water table can be defined as the limit between the saturated and unsaturated layers of earth. Water tables vary largely by location but also over time depending on water flows, e.g. fluctuating naturally with seasonal rainfall, and decreasing over time with intensive pumping for irrigation.

There are generally two types of aquifers: confined and unconfined. Confined aquifers underlie a geologic unit of low permeability which can allow water pressure in the aquifer to exceed the elevation of the low permeability confining unit (thus allowing water in a well that perforates the confining unit to rise to a greater elevation, creating an artesian effect which can sometimes result in free-flowing water at the surface). Unconfined aquifers do not feature a confining unit between the aquifer and the land surface, thus the upper limit of the aquifer is represented by the water table. In many areas, confined aquifers can underlie unconfined aquifers and groundwater basins in many parts of the world comprise multiple layers of confined aquifers separated by individual low permeability units (often referred to as aquitards). Both types of aquifer can comprise of either unconsolidated material (e.g. loose sands and/or gravel, etc.) or consolidated material (e.g. permeable/fractured sedimentary, igneous or metamorphic rock material). Unconfined aquifers generally occur at shallower depths and often feature a more direct hydraulic connection to surface water bodies but also often of smaller areal extent than confined aquifer systems that is more prone to contamination from land use activities. Deeper, confined aquifers may be more expensive to exploit.

Beyond the structure of the aquifer, the fundamental characteristics of aquifers are the stocks and flows they facilitate (Foster et al., 2013). Two properties are common to all aquifers. First, just like in the case of surface water, gravity is the main force moving groundwater from continents to water courses and oceans. Second, all aquifers have natural inflows and outflows of water, but the rates and speed of recharge and discharge vary greatly depending on local characteristics (topography, soil profile, geology, etc.). As shown in Figure 1.6, for similar local characteristics, short-distance flows going through shallow unconfined aquifers tend to evolve much faster than deeper long-distance ones involving confined aquifers.

Figure 1.6. Schematic representation of groundwater discharge and recharge

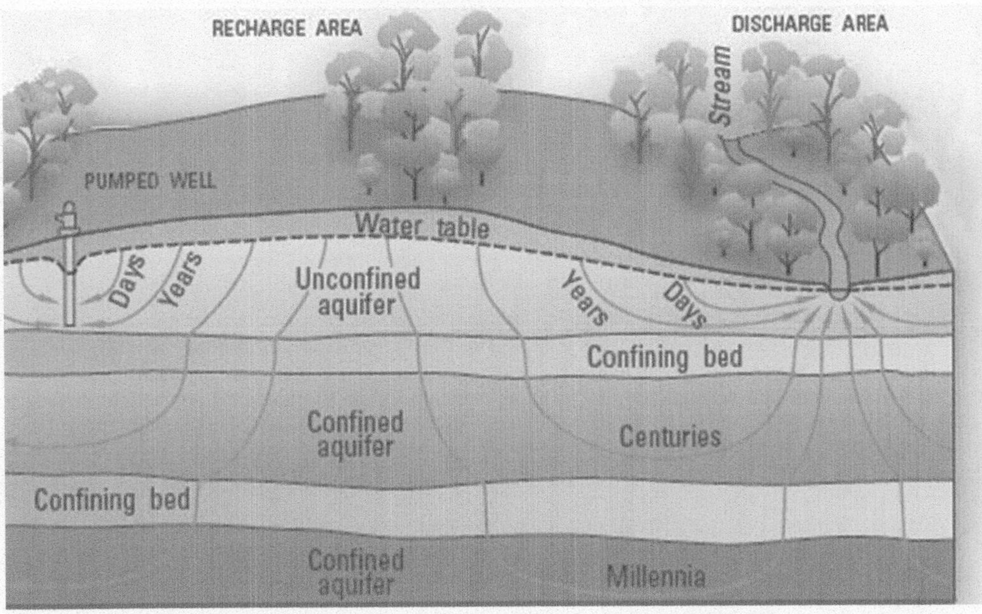

Source: USGS Water Science School, Groundwater Discharge - The Water Cycle, water.usgs.gov/edu/watercyclegwdischarge.html.

A third consideration is the renewable status of specific groundwater resources. The value of such characteristic has to be interpreted in relative term, and remains subject to discussion. All aquifers could be renewed at least partially with sufficient time under current climatic conditions, but the time it takes to renew groundwater can vary dramatically from days to hundreds of millennia. In this context, groundwater located in a large majority of aquifers, including all those used by farmers in OECD countries, can be considered renewable. Most pure non-renewable aquifers[1] are large deep and confined aquifers located in North Africa and in the Arabic Peninsula (Figure 1.7). These groundwater bodies are also called "fossil aquifers": they were formed in the geological past and do not receive any significant amount of recharge (Margat and van der Gun, 2013).

Box 1.1. Basic elements of groundwater flows *(continued)*

Naturally, this does not prevent a number of renewable aquifers in OECD countries and elsewhere to be under significant depletion. But unlike the non-renewable ones, slowing or stopping their use could result in the gradual return of reserves, while the use, even at a minimal rate, of non-renewable aquifers can be assimilated to irreversible mining, as in the case of minerals or fossil fuels.

Figure 1.7. Global distribution of groundwater resources

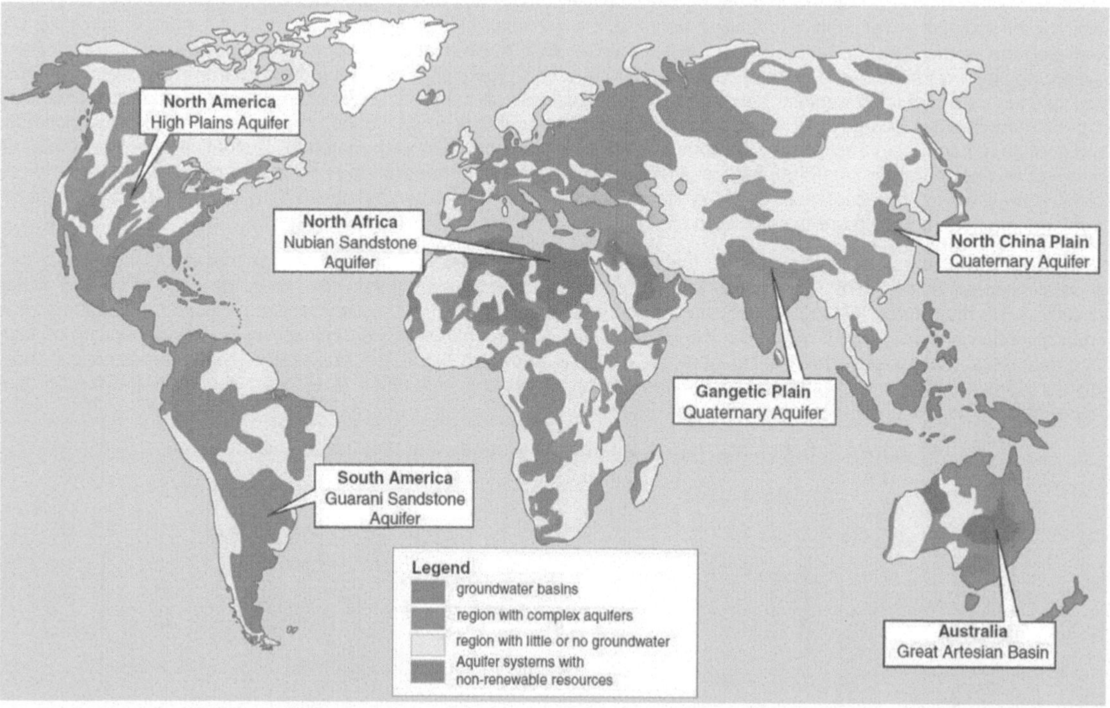

Source: Stephen Foster, John Chilton, Geert-Jan Nijsten, Andrea Richts (2013), "Groundwater — a global focus on the 'local resource'", in *Current Opinion in Environmental Sustainability*, Volume 5, Issue 6, December 2013.

Lastly, groundwater can be extracted in multiple manners, generally split into gravity based and energy based abstraction means (Margat and van der Gun, 2013). In the first case, no energy is required for the abstraction work. Examples include drains, artesian wells, infiltration galleries and underground dams, all of which may not be feasible depending on the situation. The second category includes more conventional means for agriculture, namely dug and drilled wells for pumping that can be adapted to a larger number of cases and bring about more flexibility for management, but require a source of energy.

1. Following Margat and van der Gun (2013), their mean renewal time (ratio of flux/reserve) is very large (1 000 to 100 000 years or more) compared to that of renewable aquifers (which may range from less than 1 to several 100 years).
Source: Author's own work, based on Foster et al. (2013); Margat and van der Gun (2013), and USGS (2014).

From groundwater use to groundwater stress

While volumes and areas are indicators of the importance of groundwater for irrigation, they do not inform on the risks of overdrafting. A simplified way of measuring groundwater resource risk is to compute groundwater development stress indicators (GDS), which are ratios of groundwater withdrawals over total diffuse recharge (natural and artificial). There are important caveats to such measures, ranging from the errors in data on both sides that affect the absolute values of ratios, to the different time it takes to withdraw and recharge, which will vary according to the aquifer characteristics and the season (e.g. Vrap and Lipponen, 2007: 101). While these indicators are more appropriate for use at the aquifer level, they provide a gauge of the overall scope of use over recharge at the national level (Margat and van der Gun, 2013: 175-6). Country level GDS are presented for 2010 in percentages of recharge for OECD countries in Figure 1.8, separated into agriculture (the share of agriculture use multiplied by GDS) and other uses.

The derived OECD average GDS amounts to 14.1%, 42% of which (6 %) is attributable to agriculture. Nine member countries exceed this level, in seven of these mainly due to agriculture use: Israel, Greece, Portugal, Turkey, Mexico, Italy and Spain. The United States has a slightly lower GDS, but its agricultural GDS (8.1%) exceeds the OECD average.[17] On the other hand, well water-endowed countries, like Canada and Northern European countries, non-surprisingly, have very low GDS. Countries in the middle, especially in Europe, tend to have relatively smaller GDS (under 10%) which are generally not primarily the result of agriculture use.

Figure 1.8. Estimated average groundwater development stress in OECD countries (2010)

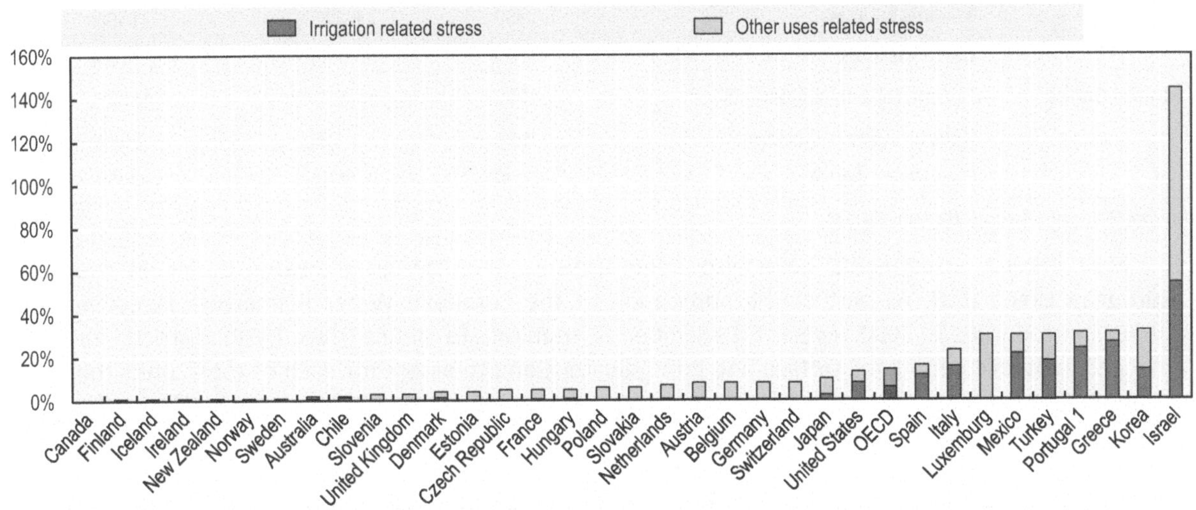

Note: the indicators are computed as the ratio of estimated aggregate national groundwater abstraction (from agriculture and other sectors) over estimated overall natural recharge multiplied by 100. It can be interpreted as the share of average recharge used for agriculture versus other uses.

1. Portugal's GDS is computed using 2013 figures based on its response 2014 OECD survey.

Source: Derived from Margat and van der Gun (2013), 2014 OECD questionnaire.

These national figures can mask significant local or sub-regional differences in stress. To demonstrate these differences, groundwater development stress indicators have been compiled at the aquifer level. Average renewable GDS values have been estimated for aquifers in multiple countries as shown (using past figures) in Table 1.3.[18] These aquifer-specific figures, often dating from a distant past, do not separate agriculture from other uses, but many of these regions have used groundwater intensively for agricultural irrigation. Within OECD, average GDS levels for the more intensively used aquifer systems in Israel, Spain, Mexico, and the United States ranged from 106% in Israel to an estimated 1022% for alluvial aquifers in Arizona.

As an alternative measurement, Gleeson et al. (2012) computed groundwater footprints with a specific focus on agricultural regions. These footprints are defined as *"the area required to sustain groundwater use and groundwater-dependent ecosystem services of a region of interest, such as an aquifer, watershed or community"*. They are computed as a modified GDS indicator — accounting for environmental flows — multiplied by the area where those are defined.[19] Within OECD countries, aquifer systems in Mexico and the United States appear to have a prominent agricultural groundwater footprint.

Table 1.3. Estimated renewable groundwater development stress (GDS) for selected aquifers in OECD countries

Country	Aquifer system	Year of estimate	Withdrawal rate (km^3/year)	Renewable GDS
Israel	Coastal aquifer	1999-2000	0.55	178%
	Mountain aquifer	1999-2000	0.76	106%
Mexico	Valley of Mexico	1998	0.63	~200%
	Baja California	1980	0.12	150%
Spain	Mancha Occidental	1989	0.58	171%
	Campo de Cartagena	1989	0.075	231%
	Sierra de Crevillente	1989	0.015	750%
	Campo de Dalias	1989	0.11	123%
	Balearic Islands	1989	0.285	109%
	Canary Islands	1989	0.34	110%
United States	Arizona (alluvial aquifers)	1990	3.78	1022%
	Central Valley of California	1990	20	286%
	High Plains Aquifer	2000	21.5	~300%

Source: Margat and van der Gun (2013).

Multi-criteria assessments have also been used to monitor groundwater basins in the European Union (EU). As part of the application of the European Water Framework Directive (WFD), introduced in 2000, every EU member is required to define river basin districts for management (accounting for surface and groundwater), and to assess for each of these units the quantitative and chemical status of groundwater resources as either "poor" or "good". On the quantitative side, several criteria have been developed to define the "good" status of a groundwater body. This includes the necessity to maintain groundwater abstraction rates under a long-term annual average, and to induce no environmental externalities (see Box 4.2 in Chapter 4). This multi-criteria definition not only focuses on stocks and flows, but also on the wider implications of groundwater use.

Yet, even with this definition, the measurement of groundwater quantity is limited and the practical methods used vary significantly from country to country. Acknowledging this limitation, Figure 1.9 shows the reported status of groundwater bodies in 19 OECD European countries with available information as of 2009. The upper panel of Figure 1.9 represents the share in the total number of groundwater bodies with good or bad quantitative status and the share of the bodies that still need to be assessed. If only 6% of the 11 897 bodies in these countries have been found to be in poor status (and 7% still undetermined), Figure 1.9 shows that this share is significant in several EU countries. The lower panel shows the absolute number of groundwater bodies under poor status in the same countries as of 2009. Of the 15 countries with "poor" groundwater bodies, only six are significant users of groundwater for agriculture (Figure 1.3).

Groundwater that flows in and out of some of the large aquifers have been studied in more detail. In the United States, the High Plains Aquifer (also called Ogallala Aquifer) and the California Central Valley aquifer systems may be among the most studied in relation to agriculture. Together, these regions accounts for about half of total groundwater depletion in the United States since 1900 (Scanlon et al., 2012). As shown in Figure 1.10, both systems have been subject to serious groundwater overdraft, with water tables lowered by up to fifty meters or more for some portions of the aquifer during the studied period. However, Figure 1.10 shows that the reduction in water levels depends on the location within the affected areas. In the case of the High Plains Aquifer, the central and southern parts of the aquifer (located in southwest Kansas and northern Texas) are hotspots for water level reduction driven by agriculture irrigation (e.g. Chaudhuri and Ale, 2014), whereas the entire northern part of the aquifer, which benefits from more natural recharge, is not much affected and has even seen increased groundwater levels locally. Similarly in California, the southern San Joaquin (SJ) and Tulare (T) counties are the main hotspots for groundwater level reduction, while the northern part of the valley, with better water endowment and surface water, is not really subject to groundwater depletion.

The effects of these reported groundwater depeletion risks on agriculture have been studied only locally in more dynamic settings found in certain hotspot regions. Steward et al. (2013) studied the Kansas portion of the Ogallala aquifer. Using a simulation model, they found that maintaining current rates of pumping without irrigation efficiency improvement would result in peak production of corn and cattle around 2040-2050, followed by a decline in production. They also showed that to sustain agricultural production at the mid-1990s level until after 2070 would require at least a 20% reduction in irrigation.

Figure 1.9. Quantitative status of groundwater bodies in selected OECD EU countries under the Water Framework Directive (2009)

Upper panel: Share of bodies with good, poor or unknown status

Lower panel: Number of bodies with poor quantitative status

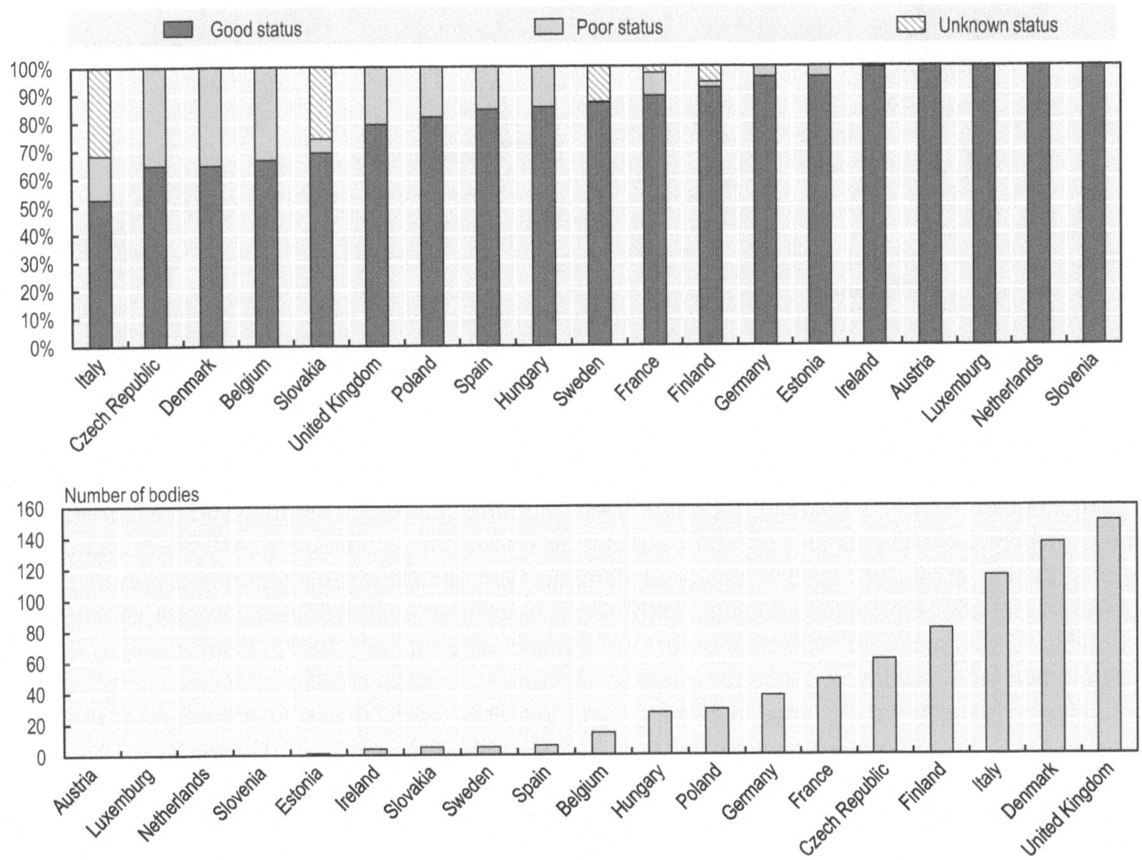

Source: EEA (2012), www.eea.europa.eu/data-and-maps/data/wise_wfd.

Figure 1.10. Changes in groundwater levels in two major US aquifer systems

Left panel (A): measured change in water tables in the High Plains Aquifer between 1950 and 2007

Right Panel (B): simulated changes in water levels in California's central valley from 1860 to 1961

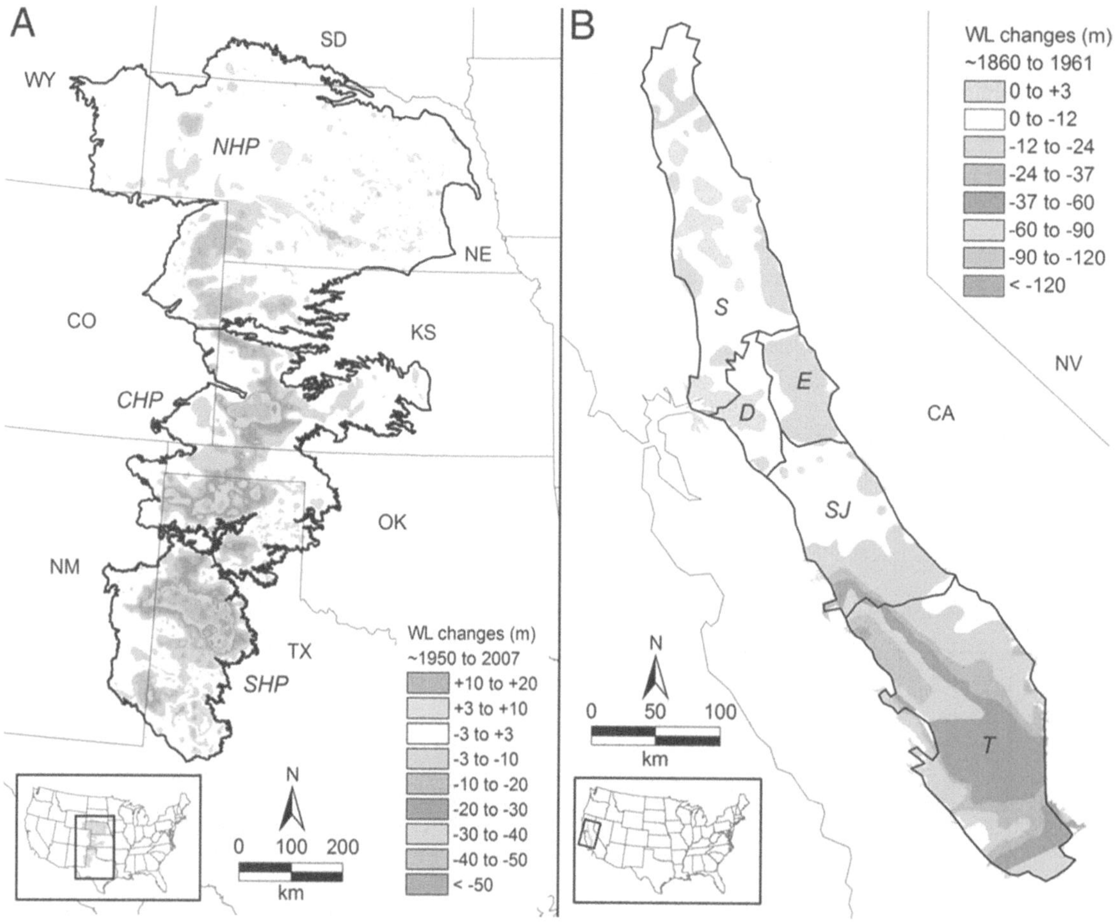

Source: Scanlon et al. (2012).

Expected effects of climate change: Increased reliance on groundwater, reduced recharge and increased salinity

Aquifers are known to respond to climate fluctuation much more slowly than surface storage, and can therefore serve as an important adaptation option for agriculture (GWP, 2012; OECD, 2014a; Wijnen et al., 2012). But increased demand with global warming may also result in increased groundwater use (Bovolo et al., 2009). To date, climate related drivers have not affected groundwater as much as non-climate drivers (Kundzewicz et al., 2007). With changes in precipitation and increased evapotranspiration, however, climate change is bound to affect groundwater directly via a change in recharge and indirectly via increases in groundwater uses (Taylor et al., 2012). These effects are expected to vary significantly regionally (Green et al., 2011); for instance, with increases (decreases) in precipitation and groundwater recharges in Northern (Southern) Europe (Hiscock et al., 2008; Negrel and Petelet-Giraud, 2011). Coupled with a rise in sea water levels, additional use in coastal areas may also lead to further salinity in groundwater (Green et al., 2011).

To a certain extent, the expected effects of climate change for agriculture can be observed in areas already subject to floods and droughts. Regions facing drought that use groundwater in conjunction with surface water irrigation substitute the latter for the former, resulting in additional extraction and use. For instance, it was estimated that groundwater volume in central California was reduced 48 times faster during the drought of 2007-09 than before (Christian-Smith and Levy, 2011). At the same time, regions where agriculture relies primarily on groundwater will draw further on the resource.[20] Coastal regions may face a higher likelihood of seawater intrusion and resulting salinity, such as in the Netherlands (de Louw, 2013). Both types may also see changes in cropping patterns and activities over time. In contrast, regions that face prolonged flooding may experience groundwater floods which would prevent most types of agriculture activities and will only make fields usable for drainage, as observed for the Chalk aquifer in Southern England (Marsh et al., 2013).

Given the lack of representative information on groundwater resources in many areas and uncertainties related to climate change, simulating the effect of climate change on groundwater irrigation is difficult (Green et al., 2011). Two types of approaches have been used to analyse how things could evolve: foresight exercises and climate-water model simulations.

Future risks for water resources were investigated by OECD (2013d in a broader study on water and climate change. In this exercise, national experts were asked to define the main water risks they envisaged, list priority quantitative and qualitative risks and key areas of vulnerability. Table 1.4 provides a summary of responses that mention explicitly groundwater quantity-related risks. It shows that OECD countries that do not use large volumes of groundwater still have significant concerns about climate change. The reduction of recharge quantity and timing are expected to be prevalent in countries like Austria, Luxemburg and Slovenia. Groundwater salinization is the other main concern, especially in countries with extensive coastal areas like Chile or Japan. Experts in Denmark and southern Europe specifically mention agriculture as an area that could be potentially affected.

A number of published studies that do not focus on climate change consistently report a likely increase in the use of groundwater for agriculture, even as overall irrigation projections do not all report such increase. Groundwater irrigation will continue to support agriculture intensification (FAO, 2003; Garrido et al., 2006). Economic drivers will increase agriculture use and pressure on aquifers especially in the Mediterranean region (Garrido and Iglesias, 2006). In Australia, the value and use of groundwater are also bound to increase given demand projections, surface water limitations, and the use of groundwater below recharge in multiple areas, with the exception of fossil aquifers (Deloitte Access Economics, 2013). Yet, global projections of irrigation tend to diverge (OECD, 2015b); the *OECD Environmental Outlook* foresees a reduced demand for irrigation in 2050, while other modelling efforts do not agree. A decline in total irrigation in parallel with a growth in groundwater irrigation would propel the share of groundwater irrigation much higher. In such a scenario, groundwater would become even more important to agriculture under climate change.

There have been several simulations of climate change effects on groundwater. These exercises generally use the amount of groundwater recharge as proxy for the impact of climate change on groundwater resources. Leterme and Maillants (2011) study the impact of climate change on groundwater and agriculture in Belgium. They find that by the year 2100 a 9% decrease in recharge is likely to occur in the Nete catchment area. Land use adaptation options suggest that a conversion of the land to maize would increase groundwater recharge and therefore reduce climate sensitivity, while a conversion to forest would lead to the opposite result. A study modelling groundwater resources projections in France under different climate change scenarios found that recharge could decrease on average by 0% up to 50% by 2070, with wider differences at the water basin level, and that there would be risks of saline intrusion in coastal aquifers (MEDDE, 2012). The IPCC (2007) report found that under a 2.4°C increase, recharge in the Ogallala aquifer in the United States would decrease by 20%. Yet, depending on the scenario and the portion of the aquifer, recharge could see an increase or decrease, and it is difficult to reach any significant conclusion (Crosbie et al., 2013). In a high resolution groundwater model focused on the Grand Forks aquifer in British Colombia, Canada, Scibek et al. (2007) show that the climate changes recharge reduction effect not only implies a decrease in groundwater levels but can also result in a significant discharge in surface water bodies (with potential impact on in-stream flow needs).

With this overview of groundwater agriculture use and trends, Chapter 2 analyses the nature and heterogeneity of groundwater irrigation systems and the challenges these face.

Table 1.4. Identified concerns for groundwater resources under climate change in OECD countries

Country	Projected impacts	Primary concern	Key vulnerability
Austria	Reduction in the duration of snow cover, decreasing groundwater recharge	Decrease in groundwater recharge	
Chile	Retreat of glaciers will have impact on groundwater	Decrease in average recharge of groundwater; and groundwater salinization in coastal zones	
Czech Republic		Decrease in groundwater level	
Denmark	Reduced formation of groundwater in summer and an increased formation the rest of the year will affect irrigation. Intrusion of seawater in groundwater		Increased demand for groundwater resources.
Estonia	Increase in groundwater recharge, depending on the hydro-geological conditions of catchments.		
Finland	Longer dry period in summer in southern Finland will reduce groundwater discharge.		
Hungary			Overexploitation of groundwater resources
Japan	Groundwater salinization due to sea-level rise.		
Korea	Depletion of groundwater		
Luxemburg	Shift in the main recharge period of groundwater.		Groundwater recharge
Mexico	Salt water intrusion in groundwater		
Netherlands	Potential salinization of groundwater resources and decrease in levels of groundwater in the summer.		
New Zealand	Reduction in groundwater supplies and higher water demand in summer		
Slovenia		Decrease in annual groundwater recharge rate	
European Union	Brackish and salt groundwater bodies will expand		Fresh water resources in Southern Europe, especially affecting agriculture

Source: OECD (2013d), http://dx.doi.org/10.1787/9789264200449-en.

Notes

1. An aquifer can be defined as "a saturated permeable geological unit (i.e. rock sediment or soil) that can transmit significant and/or economic quantities of water" (Freeze and Cherry, 1979).
2. Consumptive use can be defined as the consumption of water without direct return into water flows, i.e. via evapotranspiration.
3. Here, intrinsic physical characteristics aim to be interpreted as in the absence of flow regulation, i.e. in a context of naturalised hydrologic system.
4. The term groundwater depletion is also used to represent the same phenomenon (e.g. OECD, 2012a).
5. Changes in groundwater level may also affect biogeochemical processes which in turn can influence ecosystems.
6. The extent of groundwater drainage for agriculture use and managing floods will not be reviewed in detail. See OECD (2015a) for more information on floods.
7. Areas under continued intensive groundwater use do not appear to be limited by quality concerns, except for salinity, even if results from the survey show that some of the OECD agricultural groundwater-using regions face quality problems.
8. As Giordano (2009) points out, this invisible resource has been the object of the silent revolution, illustrating the difference in perspective between farm level resource extractions and larger scale diagnostics.
9 In Spain, for instance, there has been no overview of groundwater resources and uses at the national level since 2000-2001 (De Stefano et al., 2013).
10. There is also a poor integration of aquifer balance estimates into surface water and watershed balances, resulting in poor accuracy of the baseflow component of the overall water balances.
11. A number of general estimates in this section rely on Margat and van der Gun (2013), mostly because of the authors' rare effort to compile consistent and comparable data on resources and use across countries with a common reference date, using and reconciling data from multiple national and international databases. Their work was originally initiated under the auspice of the UNESCO-IHP program; see Margat (2008) and van der Gun (2012).
12. This includes researchers from the International Water Management Institute, the International Food Policy Research Institute, the Pacific Institute, the UNESCO-IHP, J. Margat and J. van der Gun, and universities (Complutense University of Madrid, Georgetown University, Oregon State University, Polytechnic University of Madrid, the University of California-Davis, and Wageningen University).
13. Margat and van der Gun (2013), who look at different sources (as explained in Annex 1.C), estimate withdrawals to be 982 km3/year as of 2010.
14. The year 2010 was used for consistency, compilation of aggregates and ratios. Official data for other years are used to derive the trends shown in Figure 1.5. For more on the method used by Margat and van der Gun (2013), see Annex 1.C.
15. The "overall use/area" ratio for irrigated agriculture in Portugal, for surface and groundwater irrigation, was estimated to be 7 300 m3/ha in 2009 (INE, 2011), while the 'groundwater

use/area' ratio calculated in a cross of data from the 2014 OECD survey data (for the year 2012) and IGRAC (2012) data (year 2010) was estimated to be 13 636 m3/ha.

16. Trends for other OECD countries, with lower agricultural groundwater uses, are presented in Annex 1.B.

17. Israel's GDS stands as an outlier, with an agricultural GDS exceeding 100%, but that total may be overvalued given the unaccounted use of water recycling into groundwater reserves. Portugal's official figure for total water stress (surface and groundwater) is 31%, of which 24% comes from agriculture (INE, 2011).

18. These figures were compiled renewable recharge rates with overall uses.

19. More specifically, Gleeson et al. (2012) compute groundwater footprint as A[C/(R-E)], where C, R and E are respectively the area with averaged annual abstraction of groundwater, groundwater recharge rates, and the groundwater contribution to environmental streamflow, all in units with dimensions of length/time. A is the area of any region of interest where C, R and E can be defined.

20. In such cases, an important distinction is with respect to short-run versus long-run adaptation. Hornbeck and Keskin (2014) used an empirical analysis to show that users of the Ogallala aquifers who became less sensitive to drought early on had switched to water intensive crops in the long run, thus losing their advantage and regaining sensitivity to drought.

References

Amores, M.J. et al. (2013), "Biodiversity impacts from salinity increase in a coastal wetland", *Environmental science & technology*, 47(12), 6384–92.

AQUASTAT (2013), "FAO's global information system on water and agriculture", www.fao.org/nr/water/aquastat/main/index.stm

Bovolo, C.I., G. Parkin and M. Sophocleous (2009), "Groundwater resources, climate and vulnerability", *Environmental Research Letters*, Vol. 4 (035001).

British Geological Survey (BGS) (2009), "Groundwater Information Sheet", London, UK.

Chaudhuri, S., and S. Ale (2014), "Long-term (1930-2010) trends in groundwater levels in Texas: Influences of soils, land cover and water use", *Science of the Total Environment*, Vol. 490, pp. 379–90. doi:10.1016/j.scitotenv.2014.05.013

Christian-Smith, J. and M. Levy (2011), "Impacts of the California Drought from 2007 to 2009", Pacific Institute, Oakland, CA, www.circleofblue.org/waternews/wp-content/uploads/2013/04/ca_drought_impacts_full_report.pdf

Cohen, M., Christian-smith, J. and J. Berggren (2013), "Water to Supply the Land: Irrigated Agriculture in the Colorado River Basin", Pacific Institute, Oakland, CA.

Council of Canadian Academies (2013), "Water and Agriculture in Canada: Towards Sustainable Management of Water Resources", CCA, Ottawa, www.scienceadvice.ca/en/assessments/completed/water-agri.aspx

Crosbie et al. (2013), "Potential climate change effects on groundwater recharge in the High Plains Aquifer, USA", *Water Resources Research*, Vol. 49, pp. 3936–3951.

Deloitte Access Economics. (2013), "Economic Value of Groundwater in Australia", Report for the National Centre on Groundwater Management, Melbourne.

de Louw, P. (2013), "Saline seepage in deltaic areas", Vrije Universiteit Amsterdam, the Netherlands. Retrieved from http://dare.ubvu.vu.nl/bitstream/handle/1871/48937/title_page.pdf?sequence=5.

De Stefano, L., et al. (2013), "Easier Said Than Done? The Establishment of Baseline Groundwater Conditions for the Implementation of the Water Framework Directive in Spain", *Water Resources Management*, Vol. 27(7), pp. 2691–2707.

DWF (Danish Water Forum)(2012), "Higher water security with groundwater", White Paper, Water Resources, October 2012.

EASAC (European Academies Science Advisory Council)(2010a), "Groundwater in the Southern Member States of the European Union", EASAC Policy Report 12, EASAC, Halle.

EASAC (European Academies Science Advisory Council) (2010b), "Groundwater in the Southern Member States of the European Union: an assessment of current knowledge and future prospects- Country report for Greece", EASAC, Brussels, Belgium. www.easac.eu/fileadmin/PDF_s/reports_statements/Greece_Groundwater_country_report.pdf

Esnault, L., et al. (2014), "Linking groundwater use and stress to specific crops using the groundwater footprint in the Central Valley and High Plains aquifer systems, USA", *Water Resources Research*, Vol. 50, pp. 1–21.

EEA (European Environmental Agency) (2012), WISE Water Framework Directive Database, Copenhagen. www.eea.europa.eu/data-and-maps/data/wise_wfd.

EUROSTAT (2013), "Statistics on the EU and candidate countries", website: epp.eurostat.ec.europa.eu/portal/page/portal/eurostat/home/.

Famiglietti, J. and M. Rodell (2013), "Water in the Balance", *Science*, Vol. 340, pp. 1300–1301.

Famiglietti et al. (2011), "Satellites measure recent rates of groundwater depletion in California's Central Valley", *Geophysical Research Letters*, Vol. 38(3).

FAO (United Nations Food and Agricultural Organization) (2011), "The State of the World's Land and Water Resources for Food and Agriculture: Managing systems at risk", Food and Agricultural Organization Publications, Rome.

FAO (2003), "Groundwater management: the search for practical approaches", *Water Reports*, Vol. 25, Food and Agricultural Organization Publications, Rome.

Foster, S., et al. (2013) "Groundwater — a global focus on the "local resource"", *Current Opinion in Environmental Sustainability*, Vol. 5(6), pp. 685–695.

Foster, S. (2008), "The use of non-renewable groundwater resources", Presentation at the International Groundwater Resource Assessment Center (IGRAC) 5th Anniversary Symposium, The World Bank, Washington DC.

Freeze, R.A. and J.A. Cherry (1979), *Groundwater*, Prenctice-Hall, Englewood Cliffs, N.J.

Fuentes, A. (2011), "Policies towards a sustainable use of water in Spain", OECD Economics Department Working Paper No. 840, OECD, Paris, France. http://www.oecd-ilibrary.org/economics/oecdeconomics- department-working-papers_18151973, accessed July 2013.

Garduño, H, and S. Foster. (2010), "Irrigated agriculture and groundwater resources-towards an integrated vision and sustainable relationship", *Water Science & Technology*, Vol. 67(6), 1165.

Garrido, A. et al. (2006), "Groundwater irrigation and its implications for water policy in semiarid countries: the Spanish experience", *Hydrogeology Journal*, Vol. 14(3), pp. 340–349.

Garrido, A and A. Iglesias (2006), "Groundwater's role in managing water scarcity in the Mediterranean Region", International Symposium on Groundwater Sustainability (ISGWAS), pp. 113–138.

Giordano, M. (2009), "Global Groundwater? Issues and Solutions", *Annual Review of Environment and Resources*, Vol. 34(1), pp. 153–178.

Giordano, M. and K. G. Villholth (eds.) (2007), *The Agricultural Groundwater Revolution: Opportunities and Threats to Development*, CABI, Oxford.

Gleeson, T. et al. (2012), "Water balance of global aquifers revealed by groundwater footprint", *Nature*, Vol. 488(7410), pp.197–200, doi:10.1038/nature11295.

Gleeson, T. and M. Cardiff (2013), "The return of groundwater quantity: a mega-scale and interdisciplinary "future of hydrogeology"?", *Hydrogeology Journal*, Vol. 21(6), pp. 1169–1171.

GWP (Global Water Partnership) (2012), "Groundwater Resources and Irrigated Agriculture", GWP, Stockholm, Sweden.

Gollehon, N. and W. Quinby (2006), "Irrigation resources and water costs", in US Department of Agriculture (USDA), *Agricultural Resources and Environmental Indicators*, 2006 Edition, EIB-16, Economic Research Service, USDA, Washington, D.C., pp. 24-32.

Green, T. R. et al. (2011), "Beneath the surface of global change: Impacts of climate change on groundwater", *Journal of Hydrology*, Vol. 405(3-4), pp. 532–560.

Hellegers, P. et al. (2001), "Dynamics of agricultural groundwater extraction", *Ecological Economics*, Vol. 37(2), pp. 303–311.

Hernandez-Mora N. et al. (2003), "Intensive Groundwater Use in Spain", in Llamas, R. and E. Custodian (eds.), *Intensive Use of Groundwater: Challenges and Opportunities*, Balkemapp, Lisse, pp 387–414.

Hiscock, K. et al. (2008), "Evaluation of future climate change impacts in Europe on potential groundwater recharge", *Geophysical Research Abstracts*, Vol. 10, 3pp.

Hornbeck, R. and P. Keskin (2014), "The Historically Evolving Impact of the Ogallala Aquifer: Agricultural Adaptation to Groundwater and Drought", *American Economic Journal: Applied Economics*, Vol. 6(1), pp. 190–219, doi:10.1257/app.6.1.190

Howitt et al. (2014), "Economic Analysis of the 2014 Drought for California Agriculture", Center for Watershed Sciences, University of California Davis, CA.

IGRAC (2012), "Global groundwater information system, Global overview", updated May 2012, IGRAC, Delfts, the Netherlands. htpp://ggmn.e-id.nl/ggmn/GlobalOverview.html.

IGRAC (2008), "GGIS Content of IGRAC's Global Groundwater Information System as accessible since the update of 2008" (IGRAC's web site currently under reconstruction: http://www.un-igrac.org/)

INE (Instituto Nacional de Estatistico) (2011), "Use of water in Agriculture: 2011", INE, Lisbon. www.ine.pt/xportal/xmain?xpid=INE&xpgid=ine_publicacoes&PUBLICACOESpub_boui=119577896&PUBLICACOESmodo=2&xlang=en.

IPCC (Intergovernmental Panel on Climate Change) (2007), "Climate Change 2007: Impacts, Adaptation and Vulnerability", in M.L. Parry, et al. (Eds.), *Fourth Assessment Report of the Intergovernmental Panel on Climate Change*, Cambridge University Press, Cambridge.

Kemper, K.E. (2007), "Instruments and institutions for groundwater management" in Giordano and Villholth (eds.), *The Agricultural Groundwater Revolution: Opportunities and Threats to Development*, CABI, Oxford.

Koundouri, P. (2004), "Current Issues in the Economics of Groundwater Resource Management", *Journal of Economic Surveys*, Vol. 18(5), pp. 703–740.

Kundzewicz, Z.W. et al. (2007), "Freshwater resources and their management", in Parry et al. (Eds.), *Climate Change 2007: Impacts, Adaptation and Vulnerability*, Cambridge University Press, Cambridge, pp. 173–210.

Leterme, B. and D. Mallants (2011), "Climate and land-use change impacts on groundwater recharge", Proceedings of the ModelCARE 2011, 1-7. Retrieved from www.pc-progress.com/documents/customer_projects/Mallants_ModelCARE2011.pdf.

Margat, J. (2011), "National freshwater withdrawal statistics (56 countries)", unpublished.

Margat, J. (2008), "Exploitations et utilisations des eaux souterraines dans le monde", UNESCO-IHP and BRGM Report, Paris and Nancy.

Margat, J. and J. van der Gun (2013), *Groundwater around the World: A Geographic Synopsis*, CRC Press/Balkema, Taylor and Francis, London.

Marsh, T. et al. (2013), "The 2010-12 drought and subsequent extensive flooding: a remarkable hydrological transformation", Met Office, London.

Ministère de l'Écologie, du Développement Durable et de l'Énergie (MEDDE) (2012), "Hydrologie Souterraine – Synthèse », Report from the Explore 2070 project, Paris.

Molinero, J.E. et al. (2011), "Groundwater in Spain: legal framework and management issues", Real Academia de Ciencias Exactas, Físicas y Naturales Madrid, upcommons.upc.edu/handle/2117/14364.

Montginoul, M. and J. Rinaudo (2013), "Quels mécanismes de régulation des prélèvements en eau souterraine? Comparaison du point de vue des agriculteurs, des institutionnels et des citoyens" *Science Eaux et Territoires*, Vol. 11, pp. 64–69.

Morris, B.L. et al. (2003), "Groundwater and its Susceptibility to Degradation: A Global Assessment of the Problem and Options for Management", UNEP, Nairobi, Kenya.

Negrel, P. and E. Petelet- Giraud (2011), "Isotopes in groundwater as indicators of climate changes", *Trends in Analytical Chemistry*, Vol. 30, pp. 1279-1290.

NGWA (National Ground Water Association) (2013), "Groundwater use for American agribusiness", NGWA, Westerville, OH. http://www.ngwa.org/PublishingImages/Fundamentals/NGWA_Fact_sheet1.pdf.

OECD (2015a), "Policy approaches to droughts and floods", [COM/TAD/CA/ENV/EPOC(2014)43/FINAL], OECD, Paris, http://www.oecd.org/officialdocuments/publicdisplaydocumentpdf/?cote=COM/TAD/CA/ENV/EPOC(2014)42/FINAL&docLanguage=En.

OECD (2015b), *Water Resources Allocation: Sharing Risks and Opportunities*, OECD Studies on Water, OECD Publishing, Paris, http://dx.doi.org/10.1787/9789264229631-en.

OECD (2014a), *Climate Change, Water and Agriculture: Towards Resilient Systems*, OECD Studies on Water, OECD Publishing, Paris, http://dx.doi/10.1787/9789264209138-en.

OECD (2014b), "Modelling adaptation to climate change in agriculture", [COM/TAD/CA/ENV/EPOC(2013)19/FINAL], OECD, Paris, http://www.oecd.org/officialdocuments/publicdisplaydocumentpdf/?cote=COM/TAD/CA/ENV/EPOC(2013)19/FINAL&docLanguage=En.

OECD (2013a) "Costs and distributional impacts of inaction", in OECD, Water Security for Better Lives, OECD Publishing, Paris, http://dx.doi.org/10.1787/9789264202405-10-en.

OECD (2013b), Making Water Reform Happen in Mexico, OECD Studies on Water, OECD Publishing, Paris, http://dx.doi.org/10.1787/9789264187894-en.

OECD (2013c), "OECD Compendium of Agri-environmental Indicators", Unpublished country questionnaires, available on request (tad@oecd.org).

OECD (2013d), *Water and Climate Change Adaptation: Policies to Navigate Uncharted Waters*, OECD Studies on Water, OECD Publishing, Paris, http://dx.doi.org/10.1787/9789264200449-en.

OECD (2013e), *Water Security for Better Lives*, OECD Studies on Water, OECD Publishing, Paris. DOI: http://dx.doi.org/10.1787/9789264202405-en.

OECD (2012a), "Environmental Outlook to 2050" [ENV/EPOC(2011)7/FINAL], OECD, Paris, http://www.oecd.org/officialdocuments/publicdisplaydocumentpdf/?cote=ENV/EPOC/WPRPW/WPEI(2011)1/FINAL&docLanguage=En.

OECD (2012b), *Meeting the Water Reform Challenge*, OECD Studies on Water, OECD Publishing, Paris, http://dx.doi.org/10.1787/9789264170001-en.

OECD (2011a), *Benefits of Investing in Water and Sanitation: An OECD Perspective*, OECD Studies on Water, OECD Publishing, Paris, http://dx.doi.org/10.1787/9789264100817-en.

OECD (2011b), *Evaluation of Agricultural Policy Reforms in Turkey*, OECD Publishing, Paris, http://dx.doi.org/10.1787/9789264113220-en.

OECD (2010), *Sustainable Management of Water Resources in Agriculture*, OECD Studies on Water, OECD Publishing, Paris, http://dx.doi.org/10.1787/9789264083578-en.

OECD (2009a), "Agriculture water pricing: United States" [COM/TAD/CA/ENV/EPOC/RD(2008)52/FINAL], OECD, Paris, http://www.oecd.org/officialdocuments/publicdisplaydocumentpdf/?cote=COM/TAD/CA/ENV/EPOC(2013)19/FINAL&docLanguage=En.

OECD (2009b), *Managing Water for All: An OECD Perspective on Pricing and Financing - Key Messages for Policy Makers*, OECD Publishing, Paris, http://dx.doi.org/10.1787/9789264060548-en.

OECD (2008), *Environmental Performance of Agriculture in OECD Countries Since 1990*, OECD Publishing, Paris, http://dx.doi.org/10.1787/9789264040854-en.

OECD (1998), "OECD Workshop on the sustainable management of water in agriculture – the Athens Workshop – Case studies" [COM/AGR/CA/ENV/EPOC(98)87], OECD, Paris, http://www.oecd.org/officialdocuments/publicdisplaydocumentpdf/?cote=COM/AGR/CA/ENV/EPOC%2898%2987&docLanguage=En.

Scanlon, B R., et al. (2012), "Groundwater depletion and sustainability of irrigation in the US High Plains and Central Valley", *Proceedings of the National Academy of Sciences of the United States of America,* Vol. 109(24), pp. 9320–5, doi:10.1073/pnas.1200311109.

Schoengold, K. and D. Zilberman (2007), "The economics of water, irrigation, and development", *Handbook of Agricultural Economics*, Vol. 3(06).

Scott, C., S. Dall'erba and R.D. Caravantes (2010), "Groundwater Rights in Mexican Agriculture: Spatial Distribution and Demographic Determinants", *The Professional Geographer*, Vol. 62(1), pp. 1–15.

Scibek, J. et al. (2007), "Groundwater-surface water interaction under scenarios of climate change using a high-resolution transient groundwater model", *Journal of Hydrology*, Vol. 333 (2-4), pp. 165-181.

Shah, T. (2008), *Taming the anarchy: groundwater governance in South Asia, Resources for the Future*, Washington, DC, and International Water Management Institute, Colombo, Sri Lanka.

Shah, T. et al. (2007), "Groundwater: a global assessment of scale and significance", in D. Molden (ed.), *Water for Food, Water for Life: A Comprehensive Assessment of Water Management* (pp. 395–423), International Water Management Institute, Sri Lanka.

Siebert, S. et al. (2010), "Groundwater use for irrigation – a global inventory", *Hydrology and Earth System Sciences*, Vol. 14(10), pp. 1863–1880.

Sneed, M., J. Brandt and M. Solt (2013), "Land Subsidence Along the Delta-Mendota Canal in the Northern Part of the San Joaquin Valley, California, 2003-10", *Scientific Investigations Report 2013–5142*, United States Department of Interior and United States Geological Survey, Washington, DC, http://pubs.usgs.gov/sir/2013/5142/.

Sophocleous, M. (2009), "Review: groundwater management practices, challenges, and innovations in the High Plains aquifer, USA—lessons and recommended actions", *Hydrogeology Journal*, Vol. 18(3), pp. 559–575, http://www.springer.com/earth+sciences+and+geography/hydrogeology/journal/10040.

Sophocleous, M. (2012), "Conserving and extending the useful life of the largest aquifer in North America: the future of the High Plains/Ogallala aquifer", *Groundwater*, Vol. 50(6), pp. 831–9.

Steward et al. (2013), "Tapping unsustainable groundwater stores for agricultural production in the High Plains Aquifer of Kansas, projections to 2110", *Proceedings of the National Academy of Sciences*, www.pnas.org/cgi/doi/10.1073/pnas.1220351110.

Struzik, E. (2013), "Underground Intelligence: The need to map, monitor, and manage Canada's groundwater resources in an era of drought and climate change", Munk School of Global Affairs, University of Toronto, Toronto, powi.ca/wp-content/uploads/2013/05/POWIUndergroundIntelligence- Struzik-June25.pdf, accessed July 2013.

Taylor, R.G. et al. (2012), "Ground water and climate change", *Nature Climate Change*, Vol. 3, pp. 322-329.

UNECE (United Nations Economic Commission for Europe) (2011), "Second Assessment of Transboundary Rivers, Lakes and Groundwaters" Executive Summary, UNECE, Geneva.

UNESCO (2008), "Groundwater Resources Assessment under the Pressures of Humanity and Climate Change (GRAPHIC): A Framework Document", GRAPHIC Series Number 2, UNESCO, Paris.

USGS (United States Geological Survey) (2014), USGS Water Science School Website, USGS, Washington, DC, ga.water.usgs.gov/edu/watercyclegwdischarge.html.

van der Gun, J. (2012), "Groundwater and Global Change: Trends, Opportunities and Challenges", United Nations World Water Assessment Programme, UNESCO, Paris. www.unesco.org/new/fileadmin/MULTIMEDIA/HQ/SC/pdf/Groundwater%20and%20Global%20Change.pdf.

Vrba, J. and A. Lipponen (eds.) (2007), "Groundwater resources sustainability indicators", *IHP-VI, Series on Groundwater*, No. 14, UNESCO-IHP, Paris.

Wada, Y., L.P.H. van Beek and M.F.P. Bierkens (2012), "Nonsustainable groundwater sustaining irrigation: A global assessment", *Water Resources Research*, Vol. 9, W00L06.

Wada, Y. et al. (2010), "Global depletion of groundwater resources", *Geophysical Research Letters*, L20402.

Wijnen et al. (2012), "Managing the Invisible: Understanding and Improving Groundwater Governance", The Water Partnership Programme, World Bank, Washington, DC.

Annex 1.A1

Agricultural activities supported by groundwater in OECD countries

This section presents additional figures on agricultural activities supported by groundwater. In the absence of maps on all activities, these are based on different countries and sources.

Groundwater use for livestock and aquaculture

Water used for livestock is often ignored given its proportion relative to field crop irrigation. However, it can be significant in some countries.

- In the United States, together with aquaculture, it represents 4% of total groundwater use for an industry worth USD 60 billion per year, exceeding any field crop contribution (NGWA, 2013). Groundwater-depending livestock is one of the pillars of the High Plains economy (Sophocleous, 2012). An estimated 15 million cattle and 4.25 million hogs depend on the aquifer (Sophocleous, 2009).
- In arid areas of Australia, groundwater provides the only source of drinking water for livestock, particularly for cattle and sheep, with a respective estimated value of AUD 393 million (Deloitte Access 2013).

Groundwater use for irrigated crops

Table 1.5 provides available data on the importance of groundwater for selected field crops in the United States, but other crops rely on groundwater irrigation, such as sorghum (for grain or seed), beans (dry edible), and other small grains; alfalfa, sugar beets, vegetables and potatoes. Esnault et al. (2014) estimate the groundwater footprints of 19 specific crops in the two most used groundwater basins in the United States, the High Plains Aquifer and the Central Valley of California. They find that hay and haylage and corn for grain represent the largest share of these footprints, thereby highlighting the role of feed in groundwater overdraft. Cotton also accounts for a significant share of the footprint in the Southern High Plains. Orchard, vineyards and nut trees are significant users, including in California. It also supports pastureland in all the main regions of the United States.

Responses from the 2014 OECD questionnaire on groundwater confirm the diversity of groundwater use in irrigation in other countries. Groundwater supports field crops, such as wheat (France), corn (France and Mexico), paddy rice (Japan), cotton (Mexico), sugarbeet (France), vegetables (France and Italy) and nurseries (Italy). But permanent crops or trees are also supported, such as olive trees (Italy) and sweet nuts (Mexico).

Table 1.A1.1. Scale of groundwater irrigation for selected field crops in the United States

Crop	Number of farms using groundwater for irrigation	Irrigated area (ha)	Share of total groundwater irrigation area
Corn	28085	4.3 million	32%
Soybeans	21340	2.6 million	19%
Wheat	9535	1.1 million	8.2%
Cotton	5451	1.1 million	8.1%
Rice	3861	0.73 million	5.4%

Source: Derived from NGWA (2013). The last column is derived using total estimated area by Margat and van der Gun (2013).

Annex 1.A2

Groundwater use: 2010 estimates and national trends in other OECD countries

Table 1.A2.1. Groundwater areas and estimated use in OECD countries as of 2010

	Area irrigated by GW (ha)*	Share irrigated area*	Groundwater abstraction for irrigation (km^3/year)**	Share of total groundwater abstraction**	Total groundwater abstraction (km^3/year)**
Australia	537 030	21.10%	2.32***	47%***	4.96
Austria	28 481	83.20%	0.046***	8%***	0.55***
Belgium	1 075	58.10%	0.026	4%	0.65
Canada	n.a	n.a.	0.084***	10.6%***	0.79***
Chili	58 900	5.40%	0.7154***	73%	0.98
Czech Republic	1 156	6.90%	0.011	3%	0.38
Denmark	201 480	100.00%	0.247***	38%	0.65
Estonia	0	0.00%	0	0%	0.33
Finland	765***	6.0%***	0.00027***	0.1%***	0.28
France	854 248	44.60%	0.7994***	14%	5.71
Germany	184 796	78.80%	0.2332	4%	5.83
Greece	622 765	47%***	3.1025	85%	3.65
Hungary	32 782	22.00%	0.0666	18%	0.37
Iceland	0	0.00%	0	0%	0.16
Ireland	0***	0.00%***	0***	0%***	0.21
Israel	88 969	49.00%	0.47***	38%***	1.25
Italy	893 565	41%***	6.968***	67%	10.4
Japan	232 143	8.90%	2.6256***	24%	10.94
Korea	49 639	5.60%	1.861***	43%***	4.31
Luxembourg	19	70.40%	0	0%	0.02
Mexico	3 000 000***	45%***	20.92***	71%***	29.45
Netherlands	36 089	58.00%	0.060***	6%***	0.992***
New Zealand	156 144	30.70%	0.48	60%	0.8
Norway	2 505	5.80%	0	0%	0.41
Poland	7 206	10.0%***	0	0%	2.59
Portugal	n.a.	n.a.	1.857***	76%***	2.43***
Slovak Republic	8 193	7.80%	0.0108***	3%	0.36
Slovenia	201	10.70%	0.0019***	1%	0.19
Spain	1 275 563	37.10%	4.104***	72%	5.7
Sweden	18 232	34.10%	0	0%	0.35
Switzerland	9 900	22.00%	0	0%	0.79
Turkey	1 729 578	49.30%	7.932***	60%	13.22
United Kingdom	53 039	39.80%	0.1944***	9%	2.16
United States	13 468 649	53%***	68.33***	62%***	109.65***
OECD total	23 768 066		124		223
OECD average		33%		56%	

Note: n.a.: Not available. Groundwater abstraction for irrigation and total groundwater abstraction data refer to 2009 for Portugal, to2011 for Netherlands and to 2011 (total) -2012(agricultural) for Canada.Korea's irrigation area includes only paddy rice and may be underestimated.

Source:* IGRAC (2012), **: Margat and van der Gun (2013) and OECD (2013) for the Czech Republic; *** 2014 OECD questionnaire.

Figure 1.A2.1. Agriculture groundwater use in other OECD countries from 1985 to 2013

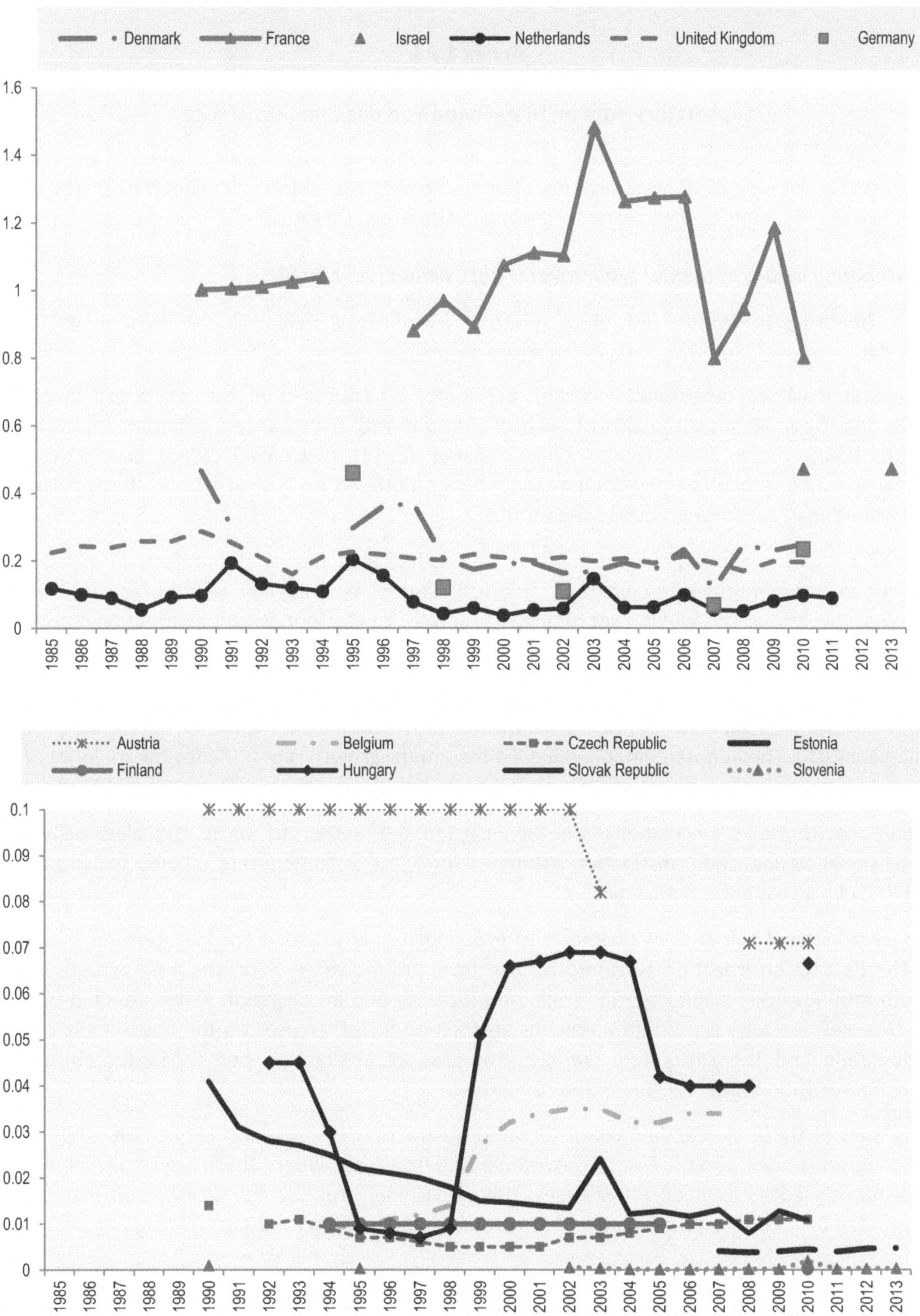

Source: OECD (2013c), 2014 OECD Questionnaire on groundwater use in agriculture and Margat and van der Gun (2013) for 2010.

Annex 1.A3

Explanatory note on Margat and Van der Gun (2013) data

This annex[1] provides an explanatory note on the country-wide aggregated values of groundwater abstraction for the year 2010, as shown in Appendix 5 of Margat and van der Gun (2013), and used as default values for countries that did not report any estimate in Chapter 1.

Estimating values of annual groundwater abstraction (year 2010)

The following procedure has been followed for deriving synchronised national abstraction estimates:

- Aggregated values of groundwater abstraction for all countries in the world are inventoried, selecting the most recent values of each of the following five (partly overlapping) data sources: IGRAC (GGIS version 2008), Margat (2008), Margat (2011), AQUASTAT (2013), EUROSTAT (2013), copying – among others – the total annual volume pumped (in km^3/year) and the corresponding reference year, as indicated by the data source.
- For each country, out of a maximum of five alternatives (see above), the preferred one is selected. In general, the most recent value is selected, unless its reliability or accuracy is considered comparatively low. This yields a set of unsynchronised "raw" groundwater abstraction data.
- Next, extrapolation is applied in order to synchronise the values and produce a provisional set of groundwater abstraction estimates valid for the year 2010. To this end, an adopted annual growth rate is used to extrapolate the raw groundwater abstraction data from their reference year up to the year 2010. The adopted default values of the annual growth rate is 0% for the majority of OECD countries, with the exceptions of Chile (1%), Israel (3%), Korea (3%), Mexico (1%) and Turkey (3%).
- Additional tentative calculations are then carried out, either to verify the plausibility of the provisional standardised abstraction estimates for 2010 or to generate a value for countries for which such an estimate is missing.
- These additional calculations estimate: (i) groundwater abstraction for irrigation on the basis of Siebert's data on irrigation consumptive use from groundwater or on the area actually irrigated from groundwater, assuming 70% irrigation efficiency and an irrigation water demand of 10 000 m3/ha, respectively; and (ii) groundwater abstraction for other uses on the basis of the country's population and the calculated average groundwater abstraction per capita for "other uses" (i.e. non-irrigation uses) for the region concerned.
- The sum of these two additionally estimated abstraction components is not only used to fill in missing values, but also to replace provisional values in cases where these are smaller than 30% of the mentioned sum. In all other cases, the provisional groundwater abstraction estimates are upgraded to become final estimates of groundwater abstraction for 2010.

1. This annex was written by Dr. Jac van der Gun (consultant, former senior hydrogeologist and director of the International Groundwater Resources Assessment Centre- IGRAC).

Estimating the break-down of annual groundwater abstraction (year 2010)

Three main groundwater use sectors are distinguished: irrigation water use, domestic water use and industrial water use. Each must be interpreted in a wide sense in order to ensure that abstractions for the three sectors add up to the total groundwater abstraction. It should be noted that available statistics do not in practice rigorously follow consistent water use class definition criteria. Rather, they follow the main targets of the groundwater withdrawal provisions, which means that rural drinking water supplied by irrigation wells may be included in "irrigation use" statistics, whereas water from public water supplies but used for small-scale irrigation or for small business or industrial uses may be included under "domestic use". Data on the break-down of groundwater abstraction with respect to water use sectors are much scarcer than data on total abstraction, and time series are virtually non-existent. In view of this, as a first approximation, it is assumed that the breakdown over the water use sectors does not vary much in time, hence even data of ten years earlier could still be representative for 2010.

The following procedure for the three main use sectors has been followed to derive best estimates of the breakdown of groundwater abstraction in 2010:

- Data on the break-down of total groundwater abstraction according to water use sectors (percentages + reference year) have been inventoried. They include IGRAC 2008 data (for 117 countries); Margat (2008) data (42 countries) and Margat (2011) data (56 countries).
- For each country for which such data are available, the preferred set of percentages is selected from possible alternatives (if available). In general, the most recent set is selected, unless its reliability or accuracy is considered comparatively low (e.g. by not adding up to 100%). This yields adopted data on the breakdown data (in %) of groundwater abstraction according to three water use sectors.
- The corresponding break-down of the groundwater abstraction for 2010 (in km3/a) could be defined in this way for 114 countries. Nevertheless, these represent together 95% of the total global groundwater abstraction. Missing data thus refer mainly to countries that abstract minor quantities of groundwater.
- From the list of 114 countries for which break-down percentages have been adopted, average break-down percentages are calculated for each of the seven regions. These percentages can be used to produce a first rough estimate for those countries in the different regions for which data are missing.
- Note: In the calculation of the total global groundwater abstraction a few (mostly small) countries have not been included. However, together these countries represent only 0.029% of the global population or 0.245% of the earth's land surface. Consequently, neglecting these countries does not significantly affect the global statistics.

Chapter 2

Understanding agricultural groundwater systems and challenges

This chapter discusses the diversity of agricultural groundwater systems in OECD countries with the goal of identifying the main factors that may need to be accounted for when managing groundwater systems. Acknowledging these characteristics, the key challenges associated with agricultural groundwater pumping in OECD countries are reviewed, considering, in particular, reversible and irreversible externalities.

The statistical data for Israel are supplied by and under the responsibility of the relevant Israeli authorities. The use of such data by the OECD is without prejudice to the status of the Golan Heights, East Jerusalem and Israeli settlements in the West Bank under the terms of international law.

Key messages

Groundwater is essentially a local resource, the characteristics of which vary greatly and depend on specific conditions and use at the aquifer level. This heterogeneity raises the question of how management challenges can be analysed and lead to meaningful responses across countries without oversimplification.

To cope with this problem and support differential management and policy responses in the OECD context, a generic characterisation of agriculture groundwater systems is proposed based on four main factors: a) agro-climatic conditions, b) relative access to and availability of surface water, c) access to, and availability of, usable groundwater resources and d) trends in groundwater use and profitability. Each of these factors can then be linked to primary and secondary variables, notably geographical, climatic, and hydrogeological considerations.

In some systems, the use of groundwater for irrigation can generate important external effects affecting both agriculture and the environment. If agriculture irrigation can induce aquifer recharge, groundwater overdraft can increase pumping costs and generate negative environmental externalities. In particular, there are significant economic consequences associated with stream depletion, salinity and land subsidence. While each of these phenomena is found in multiple OECD countries, they are associated with specific groundwater irrigation systems.

The continued use of aquifers that are under pressure and facing significant environmental issues, raises questions about current management practices and has important implications for the future of groundwater and irrigated agriculture. Key questions relate to the economic incentives associated with groundwater-based farming and the potential role of public policy around groundwater management. Both topics (incentives and policies) will be addressed in Chapter 3.

A need to move beyond the wide heterogeneity in agricultural groundwater systems

Although groundwater accounts for the largest share of available freshwater, and plays a major role in agriculture globally, it is also basically a locally-specific resource (Campana, 2014). There is a large heterogeneity in the hydrogeological nature of aquifer systems at the global scale, which, combined with diverse agro-climatic conditions, production patterns and practices, translates into multiple types of groundwater irrigation systems.

Because of this heterogeneity, it is difficult to make valid judgments on agriculture groundwater management on a national or international scale. As noted in an early study on agriculture groundwater management (Snyder, 1955: vii): "*The economic implications of ground water hydrology and ground water law are best developed through detailed studies of the experience in selected ground water basins*". Indeed, the United Nations Food and Agriculture Organization (FAO) once questioned the usefulness of developing a global picture of groundwater resources given the local emphasis of its challenges (Giordano, 2009). If each aquifer-agricultural combination differs from the next, not much could be said in general about problems, and even less about their management.

Nevertheless, increased knowledge of hydrogeological conditions, similarities in groundwater pumping patterns and technologies, and the multiplication of national, regional and local case studies have made the exercise increasingly more feasible. Multiple international projects have been conducted with the goal of characterising and assessing groundwater resources at the global scale (e.g. see van der Gun, 2007). The United Nations Economic, Social and Cultural Organization's International Hydrological Program (UNESCO-IHP), the Global Water Partnership, and the common platform set up by the International Groundwater Resources Assessment Centre (IGRAC), among other programs, underline the benefits of trying to have a comprehensive overview for local cases.[1] Furthermore, projects led under the World Bank's Groundwater-Management Advisory Team (GW-MATE) have studied the use of groundwater in agriculture in several developing countries, finding some relatively generalizable cross-country conclusions (Foster and Garduño, 2013).

The first objective of this chapter is to propose a consistent, operational characterisation of groundwater irrigation systems in OECD countries for use when considering management and policy options.[2] More specifically, this chapter reviews relevant typologies in the literature, discusses what criteria stand out from others and could be used to group similar types of constraints, and uses these two steps to move towards a characterisation for groundwater irrigation systems in OECD countries.

The second objective is to provide an overview of the critical implications and challenges associated with groundwater use for irrigation, taking the characterisation as a basis for differentiation. Several types of externalities will be presented. These challenges will then serve as reference in the following chapter, moving towards necessary policies.

Characterising agriculture groundwater systems in OECD countries

Existing aquifer typologies

There have been multiple efforts to categorise aquifer systems, accounting for dimensions related to hydrogeology, geography, as well social, institutional and economic considerations. Each of these efforts attempted to address the same dilemma of trying to provide a representative framework of a large diversity of aquifers. As the late agriculture economist S. von Cyriacy-Wantrup wrote: *"In the economics of ground water, special caution is indicated when the attempt is made to generalise. On the other hand, generalising is a necessary part of the tools and the objectives of research"* (Snyder, 1955). This section rapidly reviews some of the main efforts undertaken in this area, from international classifications of aquifer systems to socio-economic typologies of groundwater irrigated agriculture.

The first characteristics of interest relate to the nature and physical properties of a given aquifer. Five main types of aquifer can be found (Box 2.1): sand and gravel, sandstone, karst, volcanic and basement aquifers (Margat and van der Gun, 2013). Each of these types is associated with specific physical properties, such as porosity, hydraulic conductivity and thickness that determine the flow and storage aquifers can allow. The two first types include the most conducive agriculture irrigation systems, and some of the most fertile land. Other types of aquifer are also significantly used for irrigation.

Box 2.1. Five main types of aquifers

- *Sand and gravel aquifers* include extensive largely used continuous aquifers (High Plains Aquifer, Central Valley California) and local alluvial valley aquifer that are present in virtually all streams. They are the most common and easily accessible, often unconfined with relatively shallow water tables.
- *Sandstone aquifers* are consolidated sand structures. They also include major aquifers and have a lower transmissivity than sand and gravel. OECD examples include the Great Artesian Basin in Australia and the Northern Great Plain in North America, but also some smaller shallow aquifers, such as the Coastal aquifer in Israel.
- *Karst aquifers* are discontinuous complex structures, formed of cavities between different rocks, are outlets for sources, have a good flow (in some cases comparable with surface streams) but heterogeneous storage capacity, and can be largely recharged by rainfall. Examples include the Chalk Aquifer in the United Kingdom and France, multiple aquifers in Greece, the Midya aquifer in Turkey, and the Yucatan Aquifer in Mexico.
- *Volcanic aquifers* are largely fragmented aquifers, often formed in fissures or porous volcanic structures. OECD examples include the Sierra Madre Occidental in Mexico, the Canary Islands in Spain, part of the Andes in Chile, and multiple aquifers in volcanic islands (Iceland, Japan).
- *Basement aquifers* are based on crystalline and metamorphic rocks, and include different structures that are not always usable. Deeper parts include discontinuous groundwater pockets with limited storage and transmission, while shallow structures can have a better storage and relatively higher transmissivity. Examples include most of Scandinavia and parts of Australia.

Source: Author's own synthesis, based on Margat and van der Gun (2013) and Bar-Or and Matzner (2010).

Still, these characteristics have to be associated with the scope and extent of aquifers to determine flows and storage potential: the degree of confinement, depth, water table elevations, and volume all matter in this regard (Box 1.1). Multiple organisations have worked together to integrate these considerations into a simplified globally applicable classification of hydrogeological settings (WHYMAP, 2004a, 2004b). Their classification includes three categories: a) major aquifers (representing 35.6% of the global coverage of aquifers), b) areas with complex hydrogeological structures (17.8%),[3] and c) areas with only local and shallow aquifers (46.6%). Their characteristics and the correspondence with geological typology of aquifers (Box 2.1) are presented in Table 2.1. Major aquifers (a) tend to have a high storage to transmissivity ratio, complex aquifer structures (b) have discontinuous water reservoirs with variable ratios, and shallow aquifers (c) include structures with much lower ratios overall. These three major classes are now used as a standard for international maps of groundwater resources (Figure 1.7).

Table 2.1. Three classes of aquifers

Hydrogeological setting	Types of aquifers	Physical characteristics	Agriculture use implications	Examples in OECD countries
Major aquifers	Mostly sand and gravel and sandstones aquifers	Significant storage, low flows	Potential for intensive irrigation use	Australia's Great Artesian Basin (see other examples in Table 1.2).
Complex hydrogeological structures	Mixed, includes karst and volcanic aquifers, and some basement aquifers	Shallow or deep, variable stock to flow ratio	Localised productive potential irrigation uses	Po valley region in Italy, Spain's main aquifers, Turkey's aquifers
Areas with local and shallow aquifers	Alluvial formations (sand and gravel), sometimes on top of basement aquifers	Limited and localised resources, higher flow than storage	Limited potential scope, as complement with surface water	Central Europe

Source: Derived from WHYMAP (2004a) and Margat and van der Gun (2013).

A third approach has been undertaken to group similar types of aquifer systems by region. The International Groundwater Resources Assessment Centre (IGRAC) has defined 36 global groundwater regions divided into 217 groundwater provinces (Margat and van der Gun, 2013). These divisions were developed focusing on the predominant characteristics of groundwater systems in continental regions. There are four main categories of groundwater regions: *basement regions* (B), *sedimentary basins* (S), *high relief folded mountains regions* (M), and volcanic *regions* (V). These categories largely match the five aquifer types presented in Box 2.1, with the two first types under the sedimentary category (S), karst aquifers under the mountain region category (M). Table 2.5 in Annex 2.A provides basic information about the sixteen groundwater regions that cover OECD countries. Most productive agricultural regions that use groundwater belong to the five regions of the (S) category (two in North America, one in Europe, one in the Middle East and one in Oceania) or to the six M categories in the set (two in North America, one in Europe, two in Asia, and one in South America).

Building on these categorisations, social scientists have moved towards a differentiation that takes into perspective the degree and intensity of use or the potential implications thereof. More specifically, three global typologies of groundwater use have been described and referred to in the agriculture literature, developed by researchers at International Water Management Institute (IWMI), the World Bank's GW-Mate project, and the FAO, respectively.

Table 2.2 provides a comparison of the main criteria and classes as well as where OECD countries would stand. Detailed information on each classification system is shown in Annex 2.A (Tables 2.6, 2.7 and 2.8). First, Shah et al. (2007) use multiple agricultural, geographical and economic variables as indicators to define four categories of countries, depending on the nature of the predominant agricultural systems and its relationship with groundwater. The four categories are used to determine the dynamic impact of intensive agricultural use. Secondly, Foster et al. (2009) separate different conditions of aquifer "exploitations", as found in various developing countries, to derive a list of three main types and nine more specific types of conditions. Their typology does not explicitly focus on agriculture use, but given the overall focus on development, clearly addresses different types of utilisation of groundwater and their consequences. Lastly, Siebert et al. (2010) reviews the role of groundwater irrigation under four conditions, depending on the type of climate and ability to withdraw groundwater from the aquifer.

The three systems fit a larger set of agro-economic conditions than those seen in most OECD countries. As shown in the last column of Table 2.2, they may not be well adapted to the groundwater irrigation systems of OECD countries. The irrigation typology of Siebert et al. (2010) may be the only one able to cover a large number of groundwater irrigation systems, even if some aquifers used for irrigation may not have a low transmission *and* low storage, or a high transmission *and* high storage.

Table 2.2. Comparing the main socio-economic typologies

	Main considerations	Key criteria or variables used	Types	Class of OECD countries
Shah et al. (2007)	Geographic, economic, and social	Groundwater irrigated area, climate, water resource, population, agriculture organisation, drivers of groundwater irrigation, economic significance of groundwater irrigation	(1). Arid agricultural systems; (2). Industrial agricultural systems; (3). Smallholder farming systems; (4).Groundwater supported extensive pastoralism	Type (2) for all countries except Turkey (Type (1)).
Foster et al. (2009)	Risk of aquifer degradation, likelihood of conflict, and level of knowledge of aquifer and groundwater conditions	In addition to the main considerations, status of development and "exploitation", quality issues, and degree of depletion	(1) At risk of extensive quasi irreversible aquifer degradation; (2) Subject to potential conflict amongst users but not at risk of quasi-irreversible aquifer degradation; (3) Insufficient (or inadequate use of) scientific knowledge to guide development policy and process	Some aquifers under (1),some others under (3).
Siebert et al. (2010)	Favourability of climatic and groundwater withdrawal conditions	Groundwater recharge (low or high), transmissivity and storage of aquifers (both low or high)	(a) Irrigation from recharge or non-renewable deep wells, (b) surface water irrigation using run-off generated in areas with favourable climatic conditions, (c) irrigation from renewable groundwater, (d) surface water irrigation	(a) Major aquifers in arid area; (b) South West United States, (c) European lowlands, (d) Western Canada

Source: Author's own synthesis, based on Foster et al. (2009), Shah et al. (2007) and Siebert et al. (2010), as presented in detail in Annex 2.A.

Lastly, two dynamic characterisations of groundwater systems provide an interesting perspective. Instead of static characteristics, they are founded on stages of the evolution in use of groundwater systems (Figure 2.1). The first one relates groundwater to surface water use for irrigation in circumstances where both are available (GWP, 2012). Surface water is seen as first predominant to becoming unsustainable. Groundwater is then used intensively, with increased popularity until reaching a peak; in the last stage both are used in a sustainable conjunctive use system (or joint surface water-groundwater management system). The second typology focuses on a schematic political economic evolution of the system (Garrido et al., 2006). Five economic variables and five main political economic stages are considered, each related to a particular era in groundwater use in specific countries. These two models do not aim to provide a perfectly accurate and detailed representation of the evolution of groundwater irrigation system, especially given the importance regional and institutional variations. Instead, they capture some of the essential steps in the evolution of these systems as observed in a number of cross-country cases.

Figure 2.1. Schematic representation of the evolution of groundwater-irrigated agriculture systems

Top panel: surface and groundwater uses over time in dual water systems
Bottom panel: socio-economic stages in the use of groundwater systems

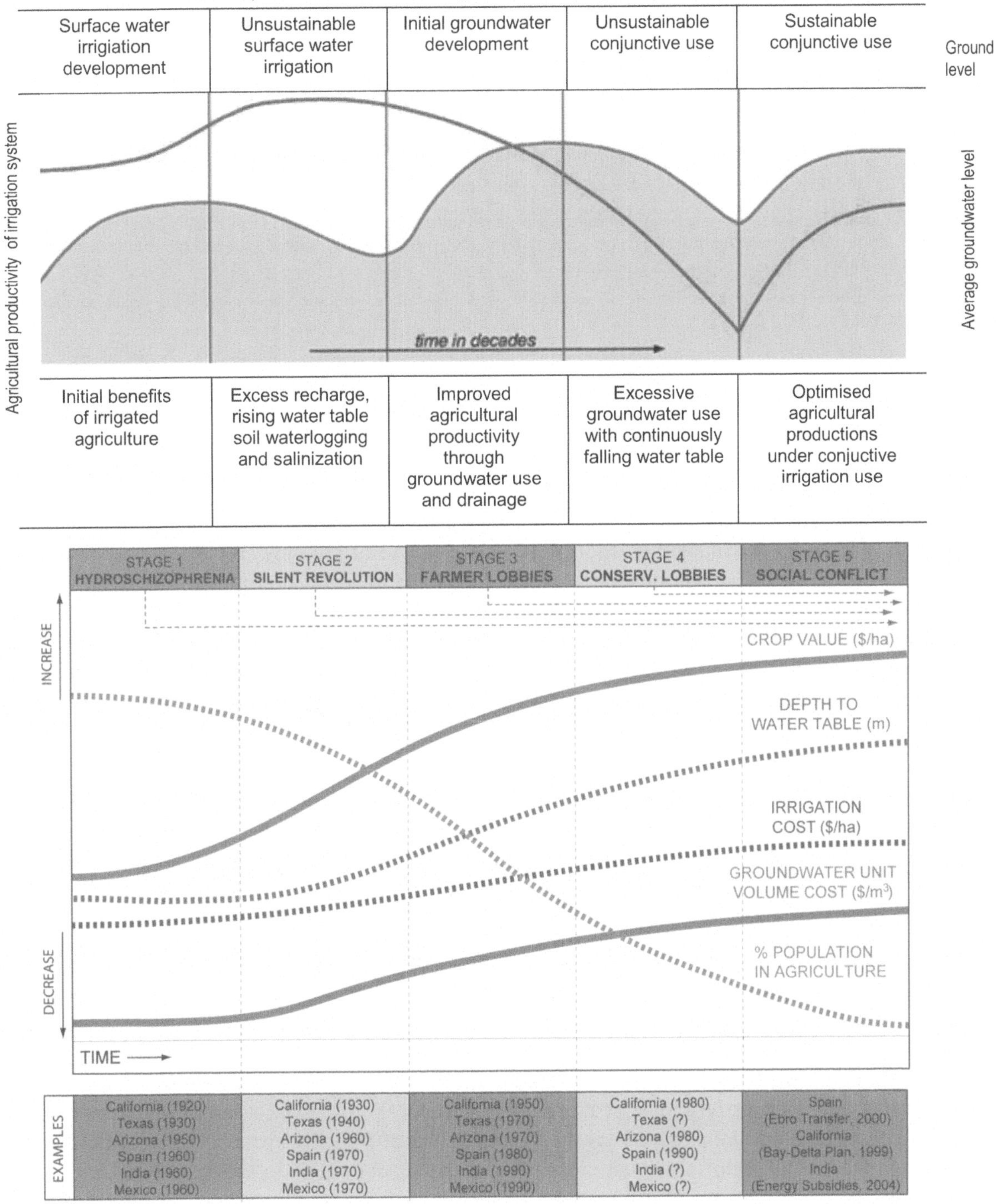

Source: Top: GWP (2012), Bottom: Garrido et al. (2006).

Main criteria of importance for OECD agriculture

There are several conditions for groundwater to be used in agriculture in OECD countries. As Shah (2008) points out (Box 2.2), climate, resources, and agriculture activities are all critical. The relationship between surface and groundwater also clearly matters, as noted in Siebert et al.'s (2010) typology. To put it simply, the following four conditions are found to be necessary for a rational but intensive use of groundwater for irrigated agriculture:[4] a) insufficient or unsteady precipitation; b) inadequate or insufficient access to surface water supply; c) accessible, available, and usable groundwater resources; and d) continued profitability of groundwater use for irrigation, especially when compared to other competing uses.

Box 2.2. Four working rules of groundwater irrigation

Shah (2008) identified the following four working rules to the intensive use of groundwater for irrigation:

"1. Intensive groundwater irrigation would emerge primarily in arid and semiarid areas that satisfy other preconditions for productive agriculture but do not have enough rainfall or surface water (such as the US great plains, Spain, or Central Mexico).

2. It would be uncommon in humid areas with abundant soil moisture and surface water (South America, Central Africa).

3. It would be uncommon in regions with poor aquifers that are costly to develop and offer low, uncertain yields often of poor quality water (Southern Africa).

4. It would decline on its own in a region as depleted aquifers become prohibitively costly to tap for irrigation or yield poor quality water (parts of US West, Saudi Arabia)."

Source: Shah (2008).

The necessary nature of these conditions is easy to demonstrate: the absence of one of these conditions suffices to eliminate the rational for intensive use of groundwater irrigation. Large rain endowment during growing seasons eliminates the need to look for alternative resources. Steady, sufficient and efficient access to surface water[5] refrains from investment in finding and accessing groundwater. The lack of access to aquifers, insufficient or unusable groundwater resources clearly impede on its use. And the rapid degradation of profitability of groundwater aquifers, or the growing competition it faces from other sectors, may prevent investment in agricultural groundwater use.

Each of these qualitative conditions can be turned into variables to characterise groundwater resources and aquifers. Once again, climate and groundwater resources matter, as do the degree of use and relationship with surface waters. The relative comparative advantage of groundwater irrigation on surface irrigation will depend on a number of factors. Box 2.3 discusses some of the main elements, considering the interface and differences between surface and groundwater irrigation systems in general. Among the exogenous factors for farmers, the comparative cost of access to one or both options matter.

Competing uses to agriculture also affect the state and evolution of groundwater resources. As shown schematically in Figure 2.2, groundwater and surface water can both supply four main sources of demand: agriculture, urban (water sanitation) and drinking water, industry and mining, and environmental flows. While each of the two sources can be used for all purposes, they are rarely used in conjunction for exactly the same type of demand. Drinking and urban water are often sourced from groundwater, while industry and agriculture more often rely on surface water.[6] In some cases, competing uses can contribute to groundwater stress (OECD, 2014a). For instance, recently developed activities in the energy sector, using hydraulic fracking, have been found to use significant amount of groundwater locally, especially in some of the water stressed agricultural irrigation areas in the United States (Freyman, 2014). Population growth, especially in urban areas, may also create more tension for resources in agriculture surrounding such an area, especially in semi-arid or arid areas.

Box 2.3. Surface and groundwater irrigation

Surface water overlaps with groundwater irrigation in several countries (e.g. Japan, see FAO, 1999), especially as buffer storage under drought (Green et al., 2011, ICID, 2010). But this is not always the case. In the United States, less than 20% of the farms and 25% of irrigated area have access to multiple water sources (OECD, 2010). Large parts of northern Mexico do not have access to surface water and therefore uses groundwater as the sole source of irrigation (Scott et al., 2010).

There are multiple physical, economic, and institutional differences between the two types of irrigation systems that condition their use as substitutes or complements.

1. *Physical access* to surface *versus* groundwater is a significant factor for irrigation. Surface water irrigation requires steady access to watercourses over time which will depend on the state and maintenance of local and river basin infrastructure, and on seasonal patterns in precipitation. Groundwater, on the other hand, is less dependent on both but in most cases requires individual investment into well-drilling and maintenance.
2. The *structure of the cost of access* to groundwater *versus* surface water may differ for users. A fixed cost is required for groundwater, and variable cost may depend on the evolution of resources and energy sources. Both fixed and variable costs are often borne by farmers who act as entrepreneurs (Garrido et al., 2006). For surface water, the cost can be largely borne by external public agencies, and variable costs depend on charges which may be subject to subsidies (Garrido et al., 2006). Surface water supplies generally require no energy at the turnout
3. While both can be considered common pool resources, *distribution and equity differ in access*. Surface water is directly and visibly dependent on river basin co-operation mechanisms, and some users have an advantage over others (i.e. upstream users). Groundwater resources do not always necessitate co-operation among users, as each operator controls his or her own pumps, but water is often available to multiple actors without visible control or cross-monitoring (Wijnen et al., 2012).
4. *Legal access and entitlements* also matter. While both may be subject to water entitlements, surface water may be managed under specific allocation systems that differ greatly from the institutional and legal frameworks that are used for groundwater. The use of Rule of Capture, for instance, whereby farmers have the right to access and use any groundwater resource under their land, is still predominant in certain parts of the United States (e.g. Peck, 2007). Under such a rule, any land owner is theoretically free to use resources on his own without restriction.
5. Managing groundwater and surface water also differs due largely to the *difference in access to information*. Regulators face a much larger asymmetry of information when considering groundwater resources than they do for surface water resources.

Building on these characteristics, wherever possible, conjunctive use of surface and groundwater can provide flexibility for farmers (Kemper, 2007), increase overall productivity (Giordano, 2009), and lower risks associated with stochastic water supply and climate volatility (Schoengold and Zilberman, 2007; Taylor et al., 2012),

Source: Author's own synthesis, based on FAO (1999), Garrido et al. (2006), Giordano (2009), Green et al. (2011), ICID (2010), Kemper (2007), OECD (2010), Peck (2007), Schoengold and Zilberman (2007), Scott et al. (2010), Taylor et al. (2012), and Wijnen et al. (2012).

Primary variables and associated elements of groundwater irrigation systems can be added to these four conditions. In particular, the following factors seem to stand out from the reviewed typologies: recharge and renewability, storage (or storativity) and transmissivity of the aquifer, stage in the use and degree of depletion. The number of users per aquifer may also matter. Indeed, taking two aquifer systems identical in terms of the other variables, but one with three users and the other with 3 000, will lead to radically different types of issues and responses (Giordano, 2009).

Figure 2.2. Groundwater and surface water uses

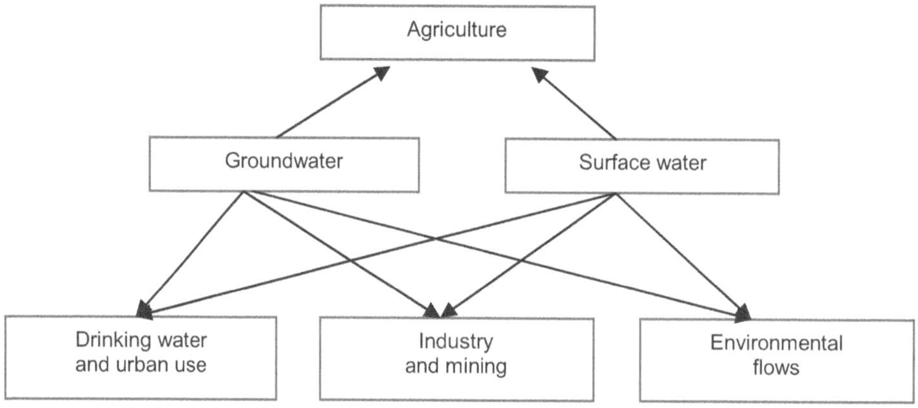

Note: Agriculture is separated from other uses for illustrative purposes.

Proposed characterisation

Four guiding principles may help to develop a usable characterisation: i) the need to ensure a comprehensive coverage, as much as possible representative of groundwater systems in OECD countries; ii) the absence of major gaps for which no case could be found; iii) the absence of main overlaps across criteria that would have the fulfilling of one criterion always imply another one; and iv) well-defined and workable boundaries that respond to the objective.

Using these four necessary conditions as the main factors as well as the discussed primary and secondary variables, a characterisation of groundwater systems is proposed in Table 2.3. The first three main factors essentially cover state variables that describe the context of any groundwater system, conditioning its potential for irrigation. The last factor covers exogenous variables related to trends in demand and overall past use which may affect the present and future potential for groundwater irrigation. These variables act as a "proxy" for the relative profitability potential of the system.

While specific cases are not presented, adopting this framework can provide a general overview of predominant groundwater systems in OECD countries. Considering the main factors, a large number of intensively used irrigation groundwater systems in OECD countries (United States, Mexico, southern Europe, Australia) tend to be in semi-arid or arid areas, some with surface water (conjunctive use) others not, with abundant and accessible but increasingly stressed groundwater resources. At the same time, looking at primary variables can help understand singularities across these aquifers, including which systems are at an advanced stage of use and those which are not, their degree of depletion, changes in competing uses, and whether they have the option to use more surface water or not. In contrast, some lesser used but still significant groundwater systems in OECD countries (e.g. in parts of central and northern Europe, Japan, Korea, Canada and Chile) may be located in relatively more humid regions, with abundant but parcelled groundwater resources used in conjunction with surface water. In these regions, agriculture may not always be the primary using sector, and they may have localised and temporary stresses. The number of users, demand drivers, cropping systems, and groundwater availability may vary significantly from one to another region.

The purpose of the characterisation is to provide elements of differentiation to be used for analysis. Current and potential groundwater use clearly depends on the four main factors and assessing how much they matter would be a useful exercise. In general, groundwater irrigation systems that are most at risk of overdraft, include those: a) with arid or semi-arid climates and water intensive crops; b) where surface water is unsteady, unavailable or insufficient; c) that have a significant access to groundwater resources with potentially limited recharge; d) that continue to experience an increase in use, either in scope (irrigation or outside) or in intensity.

The proposed system could also be used to assess the effects of changes in the main factors. Climate change may affect surface water and, as such, the likely use of groundwater in multiple areas (e.g. Famiglietti et al., 2011; chapter 1), even if it will not always result in significant short-term changes in the state of resources. But a change in surface water availability, due to infrastructure development or institutional change, is also possible and may affect the use of resources. An external shock on groundwater infrastructure (earthquake) could affect potential use. A rapid increase in the scope and use of groundwater, due to competitive demand for instance, would clearly impact the state and potential of the resource.

Table 2.3. Proposed characterisation of agriculture groundwater irrigation systems

Main factors	Primary variables of interest	Secondary variables	Examples of characteristics for OECD agricultural groundwater systems
Current and future agro-climatic conditions	Precipitation in the growing season and climate change perspectives	Cropping systems	Arid, semi-arid or humid regions
Access to surface water irrigation systems	Availability and relative cost of surface water	State of surface water Infrastructures, location in water basins.	Available and easily accessible surface water or unavailability of surface water
Availability of accessible and usable groundwater resources	Transmissivity and storage capacity; state of resource and recharge; topography and type of landscape; and quality concerns	Geological aquifer type, hydrogeological and geographic setting, depth and degree of confinement; withdrawal rate; proximity to rivers, lake or oceans.	Accessible aquifer with high storage low transmissivity and recharge rate Accessible groundwater resources from a shallow aquifer with low reserve in coastal areas. Limited accessible resources with a deep confined aquifer in a complex hydrogeological structure in a mountainous context. Low quality coastal aquifer (brackish water).
Trend in use and profitability of groundwater irrigation relative to other uses	Cost of access to groundwater	Fixed and exogenous variable costs	Declining groundwater irrigation use due to depleting stocks
	Scope of irrigation use	Number of users and irrigation area	Continued groundwater use with growing outside demand
	Intensity of irrigation use	Overall trend in use and GDS; Stage of resource use and stress	Increased use and expansion for irrigation.
	Trend in competing uses	Population growth, trend in water demand	

Key implications of groundwater use in agriculture

To complement the characterisation, this section looks at the main challenges and implications related to the use of groundwater irrigation. Three types of evolutions can be seen in intensively used groundwater resources: steady use, progressive overdraft, and mining. The consequences of such strategies on water tables are shown in Figure 2.3. While these are easy to define and difficult to measure, the question of which strategy may be best is subject to discussion.

Figure 2.3. Patterns of groundwater abstraction

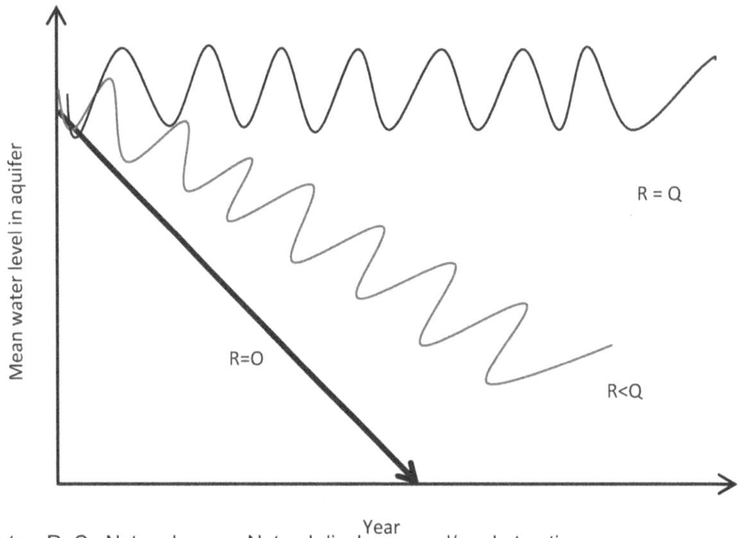

Notes: R=Q : Net recharge = Natural discharge and/or abstraction
R<Q : Natural discharge and abstraction exceed net recharge
R=O : Abstraction in absence of recharge (arid zone situation
Source: BGS (2009), http://www.bgs.ac.uk/.

Just like the definition of "renewable" may vary among practitioners (e.g. Box 1.1), the term "overexploitation" is also employed in different settings, depending on the definition of what constitutes a normal or acceptable "exploitation" path. The literature often refers to sustainable yield (or no overdraft) as the reference for "acceptable exploitation", but many authors argue that this definition does not make sense economically. Indeed, mining groundwater in non-renewable aquifers to generate capital and prepare the future can be better than keeping the stock as such (GWP, 2012). To some extent, overdrafting aquifers may lead to tremendous gains for farmers and communities by later increasing their capacity to adapt to future water constraints. As an alternative, Llamas and Garrido (2007) suggest the following definition: "*an aquifer is overexploited when the economic, social and environmental costs that derive from a certain level of groundwater abstraction are greater than its benefits*." This would imply considering a system in a dynamic cost-benefit analysis, which has some merit but also faces challenges. In practice, water management bodies define quantitative reference states to which they compare groundwater levels (EEA, 2013). Some countries even define multiple water table threshold levels for intervention (Séguin, 2009).

The underlying question around these concepts is at what point groundwater use intensity leads to unwanted consequences. This limit is reached when groundwater use generates negative externalities. Two main types of externalities can be observed in groundwater irrigation systems: extraction cost externalities and environmental externalities (Esteban and Dinar, 2012). Table 2.4 presents these effects in more detail according to their degree of reversibility. Figure 2.4 shows the percentage of agricultural groundwater regions reported in the OECD questionnaire with one or more of these externalities.

Table 2.4. Main reversible and irreversible consequences of intensive groundwater abstraction

Type	Consequences of intensive abstraction	Factors affecting susceptibility
Reversible	Pumping lifts/costs increase Borehole yield reduction Springflow/river baseflow reduction	Aquifer response characteristic drawdown below productive horizon Aquifer storage characteristics
Reversible or irreversible	Vegetative stress (natural and agricultural) Ingress of polluted water (from other aquifer or river)	Depth to groundwater table Proximity of polluted water
Irreversible	Saline water intrusion Aquifer compaction/transmissivity reduction Land subsidence and related impacts	Proximity of saline water Aquifer compressibility Vertical compressibility of overlying/inter-bedded aquitards

Source: Foster et al. (2013).

Figure 2.4. Proportion of responding regions withdrawing groundwater for agriculture in OECD countries, with at least one externality

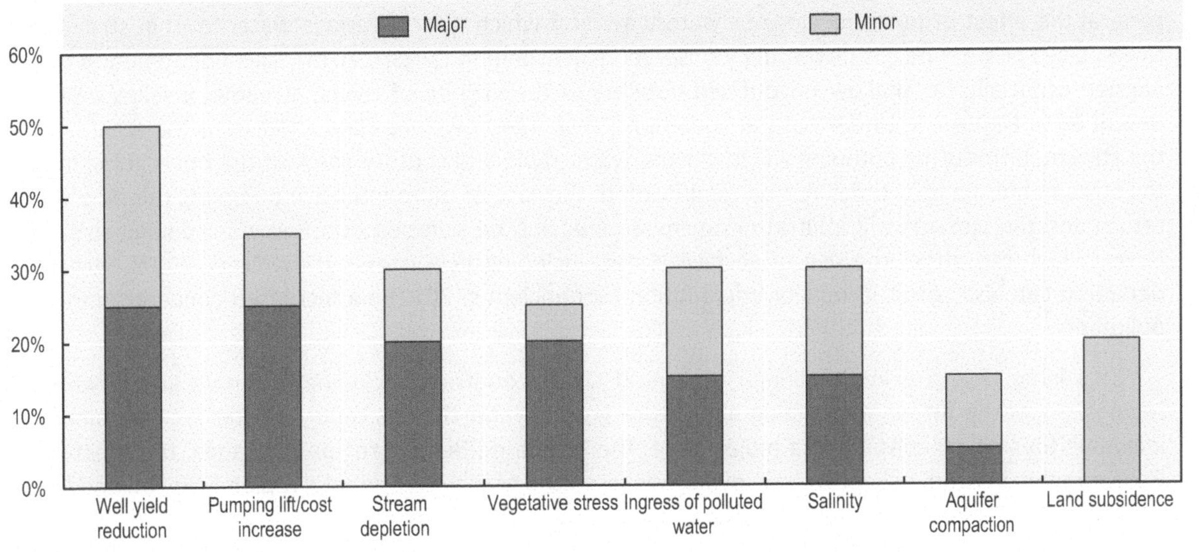

Source: Derived from the 2014 OECD questionnaire.

Extraction cost externalities

The first category of externalities involves reduced well yields and increasing pumping costs and is the most common in the responding regions. For instance, it is considered a major and growing issue in Western Galilee in Israel and Laguna Region in Mexico (2014 OECD questionnaire). As a result of sustained groundwater pumping, saturated thicknesses and well yields can be reduced significantly. In some cases, reductions in aquifer viability have been severe enough that there has been a gradual transition from irrigated back to dryland agriculture (e.g. parts of western Kansas in the United States). In other areas, such as north Texas, irrigation is still possible but it is clear that the current intensity (in terms of irrigated area and application rates) cannot be maintained in the future.

Even if reversible, such pumping can result in increased costs and lower volumes for all. Its extent clearly depends on the nature of aquifers, and may depend on the level of knowledge of users on the state of the resource (Saak and Peterson, 2007). In the proposed characterisation, pumping externalities are more problematic in groundwater irrigation systems that have an arid climate, limited surface water access, accessible resources in continuous large area aquifers with limited flows and storage but multiple users.[7]

In the case of environmental externalities, multiple consequences can arise, some reversible and others irreversible, some directly affecting agriculture activities while others only directly affecting the environment. Vegetative stress can affect all plants, while pollution ingression may depend on other activities and will not always have visible consequences on crops, depending on concentration and type, but can affect drinking water sources.

Environmental externalities

The following sub-sections will briefly review in more detail three main negative environmental externalities that can create tremendous problems for agriculture and/or the environment: stream depletion (surface- water groundwater interaction), salinization, and land subsidence.

Stream depletion (surface water-groundwater interaction)

Stream depletion, a specific type of surface water groundwater interaction which involves in general the effect of pumping close to waterways and which affects surface water level in streams, rivers, lakes or wetlands, is increasingly observed (Gleeson and Cardiff, 2013). This phenomenon can happen especially for shallow unconfined aquifers in the vicinity of rivers, streams or lakes and is described in Figure 2.5. Under natural conditions, recharge flows are conducted from the aquifer to the stream. Introducing pumping will lower the water table and capture some of the flows going into the stream, but the effect may not be significant. In the case of more intensive pumping, the flow will return and the streams will infiltrate into the aquifer and be pumped out. Beyond reducing stream flows, which can affect the use of surface water, including ecosystems (EUWIMed, 2007), stream depletion can also result in quality degradation (Sophocleous, 2012) via increased concentration of pollutants.

This issue is a growing challenge in several OECD countries, including intensive groundwater regions. Six of the fifteen responding agriculture groundwater regions report having such problems, including four which consider it a major issue: the Nappe de Beauce region of France, the Western Galilee region in Israel (where it is a growing concern), the Mancha Occidental region in Spain, and the Northern High Plains Aquifer region in the United States.

In the United States, this phenomenon has occurred widely over the High Plains Aquifer (e.g. Sophocleous, 2012). In Nebraska, while it has been estimated that only 1% of groundwater storage has been depleted, models have shown that groundwater pumping has reduced flows to the Platte River and other rivers by up to 50% (Scanlon et al., 2012). Stream depletion has led to intra- and inter-state conflicts (Kuwayama and Brozović, 2013) and can have implications on irrigation systems, reducing options for farmers. The main type of affected groundwater irrigation systems are those with conjunctive use, where groundwater is in connection and in proximity with surface water bodies.

Figure 2.5. Schematic representation of surface water groundwater interaction

A: Natural conditions; B: Moderate pumping; C: Intensive pumping

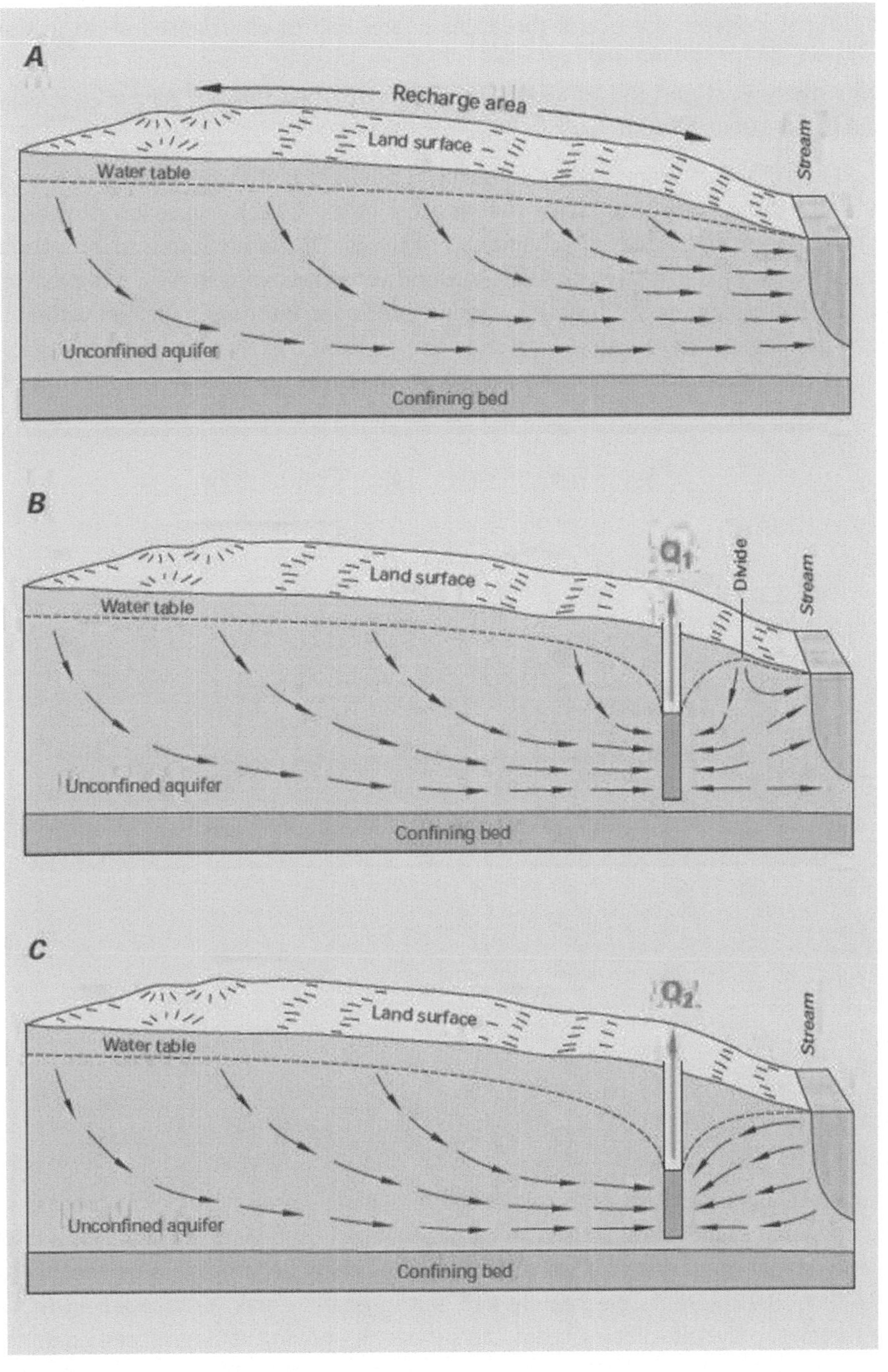

Source: Winter et al. (1998). http://pubs.usgs.gov/circ/circ1139/pdf/circ1139.pdf.

Groundwater salinity

Salinity[8] is one of the most important and growing constraints for irrigated agriculture. It may affect up to 20% of irrigated areas and threatens nearly half of all irrigated areas in the long term (Le Kama and Tomini, 2012). One and half million hectares may be taken out of production as a result of land salinity every year and the total global costs for producers may exceed USD 11 billion/year (Schoengold and Zilberman, 2007).

Even if a large share of saline water intrusion in aquifers is due to the use of groundwater for irrigation, it is not the only driving factor (Balderacchi et al., 2012). Figure 2.6 provides a synthetic representation of the main sources of groundwater salinity. Of the identified seven sources, four are directly caused by pumping: inflow of saline groundwater following heavy pumping (cone), deep intrusion of saline groundwater,[9] phreatic evaporation under intensive (surface water) irrigation in the absence of drainage (water logging), and sea water intrusion in coastal areas. A fifth category (soil fractionation) is related to irrigation of the soil, but does not necessarily result from groundwater pumping.

Figure 2.6. Main sources of groundwater salinity

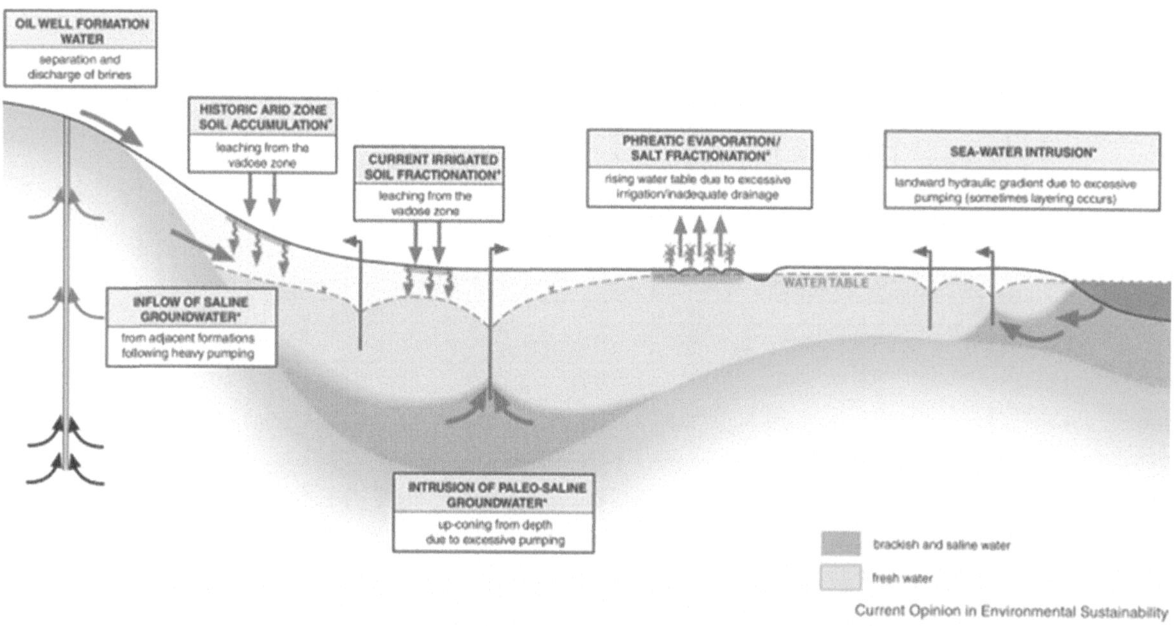

Source: Foster et al. (2013). http://ac.els-cdn.com/S1877343513001401/1-s2.0-S1877343513001401-main.pdf?_tid=08fdf794-f320-11e4-a189-00000aacb35f&acdnat=1430828144_5983e91f0c30ec60688eb04b46dfb891.

Salinization can be disastrous for agricultural activities. The physiological process is relatively straightforward: when saline water is used by plants, water is used by plants and salt is left in the root zone, therefore rapidly rendering soils saline which increasingly becomes less permeable to water (Le Kama and Tomini, 2012). Plants die rapidly and salt crystals remain embedded in the soil. In the absence of remedies, only salt tolerant crops can survive. Removing salt in soil can be done via proper drainage, but is rarely completely eliminated, and can be highly onerous (Foster et al., 2013). Prevention, using artificial groundwater recharge can work, as shown in Tunisia (Garrido and Iglesias, 2006) or using barriers as in Israel (Margat and van der Gut, 2013). The use of drip irrigation can also help slow the process (Cooley et al., 2009).[10]

A number of OECD countries are especially concerned with seawater intrusion in coastal areas (such as Greece, see EACSAC, 2010a; and the Italian plains, see EASAC, 2010b and Napoli and Vanino, 2011). In such situations, the challenges are not only to avoid intensive pumping but to keep a sufficiently high level of freshwater in aquifers and to slow the penetration of salt water. In the central coast of California, a highly productive area for produce, freshwater sources are not only sought as an alternative supply to groundwater use but also to be used as groundwater recharge to sustain aquifers in the future (Levy and Christian-Smith, 2011). In some cases, as in the Nueva Lagoon region in Spain, sea water intrusion can also filter into wetlands, affecting plants and other species in the local ecosystems (Amores et al., 2013). In other cases, with low altitude surface coastal areas, like in the Netherlands, sea intrusion in groundwater can result in seepage in surface water, affecting lakes, rivers, and entire regional water systems (de Louw, 2013).

Following the proposed characterisation, the most affected systems are those that are most inclined to use groundwater significantly (as defined above), but also those in proximity with salted water resources. This includes groundwater systems in coastal areas with limited supply in fresh surface water and a relatively high intensity of use, such as the Hermosillo coastal aquifer in Mexico (Custodio, 2003) and multiple areas in Greece (EASAC, 2010b). Other areas concerned are those where groundwater levels are in proximity with surface water (lowland regions of Europe, see Annex 2.C). Areas in proximity to underground saline water, including shallow alluvial aquifers such as those in New Zealand and paleo-channels in Western Australia, are also subject to saline constraints (Magat and van der Gun, 2013). Salinity in such areas will furthermore likely increase with climate change (Green et al., 2011).

Land subsidence

Land subsidence is another very important possible result of intensive groundwater abstraction. Drawing water in aquifers made of unconsolidated and porous geological structures, including sedimentary complexes, can result in significant compaction of aquifers that in some cases result in the sinking elevation of the land surface (Margat and van der Gun, 2013). Multiple impacts can then be observed from the deterioration of infrastructure, buildings, or even pumping systems, to the displacement of water courses and energy networks, the destruction of trees, erosion, and so on. Total damages can be extensive; in California and Texas, they were estimated to exceed USD 100 million/year (OECD, 2013).

Box 2.4. Groundwater withdrawal induced land subsidence in OECD countries

UNESCO (1984) inventoried international cases of land subsidence related to the intensive use of groundwater. They found 42 cases, mainly located in urban agglomerations, perhaps because of better measurement. More recent efforts have included dozens of new cases. Notable examples in OECD countries include:

- Italy: Milan (subsidence of 0.2 m from 1952 to 1972); Venice (more than 0.2 m since the 1930s); and the Po Delta (more than 3 m during the 1950s).
- Mexico: Mexico City (since the 1920s up to 0.4 m per year in the Centre, up to 10 m and 300 mm/year during 2004–2006); Toluca Valley (90 mm/year during 2003–2008).
- United States: Denver, Houston, Las Vegas, San Francisco, Tucson; up to 2–9 m of subsidence in several cities in California and the San Joaquin Valley; Coachella Valley, California (70 mm/year during 2003–2009); the Bolsón del Hueco basin around El Paso in Texas (0.3 m of subsidence since the 1950s).
- Japan: Tokyo (starting in 1910; subsidence up to 4 m; land surface fell to 1 m below sea level); Osaka (up to 2.5 m); the Sagamigawa alluvial plain (up to 0.32m during 1975–1995); and another 62 cases reported in 1998.

Source: Margat and van der Gun (2013), Famiglietti et al. (2011), UNESCO (1984).

Land subsidence takes place in multiple contexts, many of which are not related to agriculture. Box 2.4 provides some of the well-known reported cases in OECD countries. Japan's major urban areas have known significant land subsidence episodes, but these events were related to intensive pumping for urban water use, not agriculture (Taniguchi et al., 2008). Similarly, land subsidence has impacted the Po delta in Italy, but it was mostly related to urban and industrial development (Teatini et al., 2006). Yet multiple examples in areas of intensive use result directly from groundwater irrigation. For instance, agriculture irrigation contributed to the large subsidence observed in Guanajuato State in Mexico (Custodio, 2003) and the Central Valley of California (FAO, 2011).

Figure 2.7. Elevation change computed from repeat geodetic surveys along the Delta-Mendota Canal (left NW-right SE)

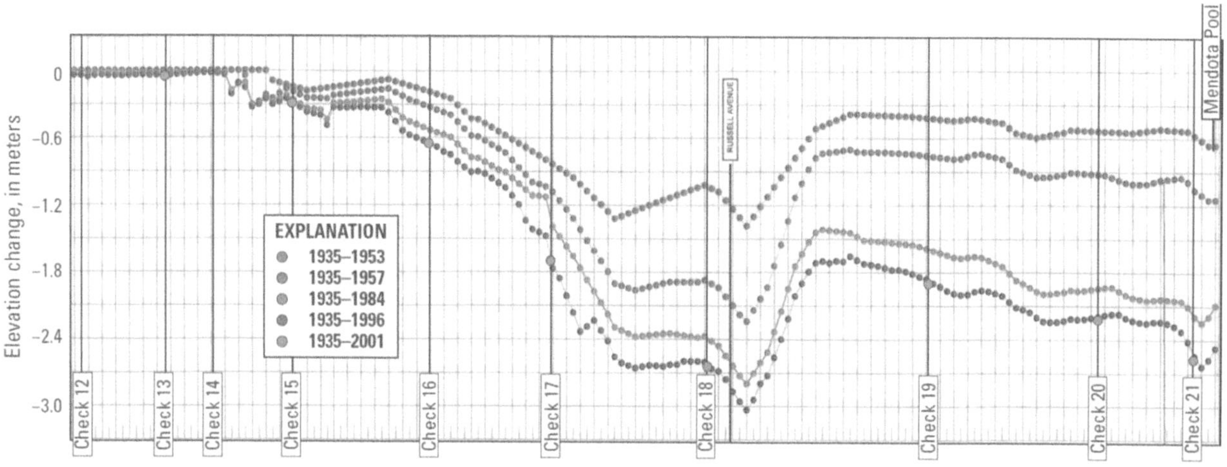

Source: Sneed et al. (2013). http://pubs.usgs.gov/sir/2013/5142/pdf/sir2013-5142.pdf.

Figure 2.7 shows the evolution of land levels in a specific area of California's central valley. In this example, land has decreased by up to 2 meters during a seventy-year span, but it was estimated to be above 8.5 meters in other areas (Sneed et al., 2013). The pattern and rhythm of land subsidence during this period has followed the evolution of water use: rapid depletion early on, then slowing down with the development of surface water irrigation infrastructure, and then again re-acceleration of the process during the more recent period of droughts (Sneed et al., 2013).

Consequences for agriculture may not always be as directly visible as those in the case of salinity, but long-term impacts can threaten any agriculture activity, damage ecosystems, discourage investments, and affect rural communities. It can also result in sea level intrusion and associated flooding, and induce water logging and salinity including in low lying delta regions (Custodio, 2003; de Louw, 2012). Groundwater pumping-induced land subsidence has even been associated with the observed seasonal uplift of the Sierra Nevada Mountains and increased seismic activities in California (Amos et al., 2014).

Prevention of such events, just like salinity, involves better management of groundwater reserves, either via a complete stop to withdrawals — a successful strategy in the case of industrial withdrawals in Venice, Italy (Margat and Van der Gut, 2013) —or via increased recharge. The use of artificial recharge systems, such as water banking in aquifers, can help (Maliva, 2014).

The main characteristics of potentially affected groundwater irrigation systems are related to the hydrogeological structure of the system and the degree of depletion of groundwater resources. Just like in the case of salinity, compactable aquifers, including those in sedimentary basins (sand and gravel) for which total withdrawal (from agriculture and potentially other sources) are significant relative to recharge, may be among the most likely to be subject to land subsidence. The sensitivity of some of the most productive aquifers to these externalities raises the question of whether their intensive use would be justified if such externalities were accounted for.

Agricultural irrigation effects on groundwater recharge

Lastly, one aspect that is not considered a challenge but rather acting as a positive externality is the fact that agricultural land use, and in particular irrigated agriculture (surface or groundwater-based), can contribute significantly to the recharge of water bodies, including groundwater (Scott and Shah, 2004). The use of any type of water in fields, mostly used for evapotranspiration, can also result in partial seepage into the soil, leading to the recharge of aquifers.

This phenomenon is well-known in multiple OECD countries. In the first part of the 20th century, the mere conversion of land to rain fed agriculture in southeast Australia and southwest United States led to significant increases in recharge and groundwater storage (Taylor et al., 2012). In central Spain, intensive groundwater pumping in the Upper Guadiana Basin has contributed to a net increase in water availability for consumptive use (Llamas and Garrido, 2007). Such a mechanism is bound to be found especially in areas with shallow unconfined aquifers with rapid recharge.

Irrigation-induced groundwater recharge is particularly significant in countries growing paddy rice (OECD, 2014b). In Japan it was estimated that irrigated rice cultivation contributes over 23% of total groundwater recharge (Mutsubishi Research Institute, 2001). Several Japanese cities have supported paddy rice cultivation in the surrounding area for recharge of aquifers, to help slow depletion and related land subsidence (see Box 4.4 in chapter 4 and OECD, 2015, for specific examples).

Interestingly, irrigation-induced recharge is even more prevalent under inefficient irrigation systems (Giordano, 2009). There is effectively a trade-off between the use of efficient irrigation systems, which allows saving water resources, and the level of recharge of aquifers they induce. For instance, GWP (2012) shows that highly efficient drip irrigation will increase potential recharge, but not as much as low-efficiency flood irrigation. Such an effect will, however, only be observed at a significant scale in relevant aquifers where irrigation-induced recharge is significant compared to natural recharge.

Yet, such recharge can also become problematic when groundwater is not sufficiently used (Margat and van der Gun, 2013). In some conjunctive irrigation systems, in particular those with unconfined aquifers, surface water irrigation will increase recharge to the point of having the water table reaching a level close to the surface, as suggested in Figure 2.1 (top panel). This can result in water logging, evaporation and salinity (Figure 2.6) and calls for proper drainage.

Notes

1. For more information on UNESCO-IHP, see http://www.unesco.org/new/en/natural-sciences/environment/water/ihp, for GWP: http://www.gwp.org/ and for IGRAC: http://www.un-igrac.org/. See van der Gun (2007) for a list of other institutions.
2. Commonalities have also been explored by the development of indicators, such as the UNESCO Groundwater Resources Sustainability Indicators (Vrba and Lipponen, 2007).
3. These are defined as "aquifers in a geological complex setting with highly productive aquifers in heterogeneous folded or faulted regions in close vicinity to non-aquifers" (WHYMAP, 2004b).

4. We exclude productive rain fed agriculture areas where irrigation is not economically justified (e.g. large parts of Canada and Northern Europe).

5. In the absence of binding allocation constraints.

6. In the European Union, for instance, groundwater represents 70% of domestic water use and only 20% of irrigation water (OECD, 2012b).

7. An additional externality not discussed here relates to the possible competition among users of the same aquifer. The extent of such behavioural phenomenon, especially in the context of large aquifers, remains subject to discussion. In particular, Pfeiffer and Lin (2013) empirically conclude that strategic behaviour is occurring in the High Plains Aquifer. But other theoretical and experimental work favours a more myopic behaviour among actors in such settings (e.g. Rubio and Casino, 2003; Gardner et al. 1997).

8. Salinity can be defined as based on the total concentration of dissolved solid (TDS). Water is defined as brackish if TDS is above 1000 mg/L and defined as saline if over 10 000mg/L (Margat and van der Gun, 2013).

9. As noted in Table 1.1, about 60% of total groundwater reserves are brackish or saline water.

10. There are also breeding programs to render traditional crops salt tolerant in a number of countries, but the results are not satisfying.

References

Amores, M.J. et al. (2013), "Biodiversity impacts from salinity increase in a coastal wetland", *Environmental Science and Technology*, Vol. 47(12), pp. 6384–92.

Amos, C.B. et al. (2014), "Uplift and seismicity driven by groundwater depletion in central California", *Nature*, Vol. 509, pp. 483-486, doi: 10.1038/nature13275.

Balderacchi et al. (2012), "Groundwater Pollution and Quality Monitoring Approaches at the European Level", *Critical Reviews in Environmental Science and Technology*, Vol. 43(4), pp. 323–408.

Bar-Or, Y. and O. Matzner (eds.) (2010), "State of the Environment in Israel, Indicators, Data, and Trends", Ministry of Environmental Protection, State of Israel, Jerusalem.

British Geological Survey (BGS) (2009), "Groundwater Information Sheet", London, http://www.bgs.ac.uk/.

Campana, M. (2014), "Groundwater Management: Quo Vadis?" In Fitch, E.J., and R. A. Engberg, (eds.), *AWRA at 50: The Future of Water Resources in the United States, Water Resources* IMPACT 16(1): 26-28, http://aquadoc.typepad.com/files/impact_gwm_quo-vadis_campana-1.pdf.

Cooley, H., J. Christian-Smith, and P. Gleick (2009), "Sustaining California Agriculture in an Uncertain Future", Pacific Institute, Oakland, CA.

Custodio, E. (2003), "Intensive groundwater development: A water cycle transformation, a social revolution and a management challenge", In L. Martinez-Cortina, A. Garrido, and E. Lopez-Gunn (eds.), *Water Workshop on Re – thinking Water and Food Security Paradigms*, CRC Press, London, http://www.rac.es/ficheros/doc/00734.pdf.

de Louw, P. (2013), "Saline seepage in deltaic areas", Vrije Universiteit Amsterdam, the Netherlands. Retrieved from http://dare.ubvu.vu.nl/bitstream/handle/1871/48937/title_page.pdf?sequence=5.

EASAC (European Academies Science Advisory Council) (2010a), "Groundwater in the Southern Member States of the European Union: an assessment of current knowledge and future prospects- Country report for Greece", EASAC, Brussels. www.easac.eu/fileadmin/PDF_s/reports_statements/Greece_Groundwater_country_report.pdf.

EASAC (European Academies Science Advisory Council) (2010b), "Groundwater in the Southern Member States of the European Union: an assessment of current knowledge and future prospects- Country report for Italy", EASAC, Brussels, http://www.easac.eu/fileadmin/PDF_s/reports_statements/Italy_Groundwater_country_report.pdf.

Esteban, E. and A. Dinar (2012), "Cooperative Management of Groundwater Resources in the Presence of Environmental Externalities", *Environmental and Resource Economics*, Vol. 54(3), pp. 443–469.

EEA (European Environmental Agency) (2013), "Assessment of cost recovery through water pricing", *EEA Technical Report*, No. 16/2013, Copenhagen.

EUWIMed (European Union Water Initiative-Mediterranean region)(2007), "Mediterranean Groundwater Report", Technical report on groundwater management in the Mediterranean and the Water Framework Directive", EUWI, Brussels.

FAO (United Nations Food and Agricultural Organization) (2011), *The State of the World's Land and Water Resources for Food and Agriculture: Managing systems at risk*, Food and Agricultural Organization Publications, Rome.

FAO (1999), *Irrigation in Asia in Figures*, Food and Agricultural Organization Publications, Rome.

Famiglietti et al. (2011), "Satellites measure recent rates of groundwater depletion in California's Central Valley", *Geophysical Research Letters*, Vol. 38(3).

Freyman, M. (2014), "Hydraulic fracturing and water stress: Water demand by the numbers. Shareholders, lenders and operators guide to water sourcing", Ceres consulting, Boston, MA.

Foster, S. et al. (2009), *Sustainable Groundwater Management Contributions to Policy Promotion*, The World Bank, Washington, DC.

Foster, S., J. Chilton, G.-J. Nijsten, and A. Richts (2013), "Groundwater — a global focus on the "local resource", *Current Opinion in Environmental Sustainability*, Vol. 5(6), pp. 685–695.

Foster, S. and H. Garduño, (2013), "Irrigated agriculture and groundwater resources-towards an integrated vision and sustainable relationship", *Water Science & Technology*, Vol. 67(6), 1165.

Gardner, R., M.R. Moore, and J.M. Walker (1997), "Governing a groundwater commons: a strategic and laboratory analysis of western water law", *Economic Inquiry* Vol. 35, pp. 218–23.

Garrido, A and A. Iglesias (2006), "Groundwater's role in managing water scarcity in the Mediterranean Region", International Symposium on Groundwater Sustainability (ISGWAS), p. 113–138.

Garrido, A., P. Martínez-Santos and M.R. Llamas (2006), "Groundwater irrigation and its implications for water policy in semiarid countries: the Spanish experience", *Hydrogeology Journal*, Vol. 14(3), pp. 340–349.

Giordano, M. (2009), "Global Groundwater? Issues and Solutions", *Annual Review of Environment and Resources*, Vol. 34(1), pp. 153–178.

Gleeson, T. and M. Cardiff (2013), "The return of groundwater quantity: a mega-scale and interdisciplinary "future of hydrogeology?", *Hydrogeology Journal*, Vol. 21(6), pp. 1169–1171.

GWP (Global Water Partnership) (2012), "Groundwater Resources and Irrigated Agriculture", GWP, Stockholm.

Green, T. R. et al. (2011), "Beneath the surface of global change: Impacts of climate change on groundwater", *Journal of Hydrology*, Vol. 405(3-4), pp. 532–560.

IGRAC (International Groundwater Resources Assessment Centre) (2004), "Short descriptions of global groundwater regions", IGRAC, Delfts, http://www.un-igrac.org/dynamics/modules/SFIL0100/view.php?fil_Id=59.

IGRAC (2012), "Global groundwater information system, Global overview", updated May 2012, IGRAC, Delfts, http://ggmn.e-id.nl/ggmn/GlobalOverview.html.

IPCC (Intergovernmental Panel on Climate Change) (2007), "Climate Change 2007: Impacts, Adaptation and Vulnerability", in M.L. Parry et al. (Eds.), *Fourth Assessment Report of the Intergovernmental Panel on Climate Change*, Cambridge University Press, Cambridge.

Kemper, K.E. (2007), "Instruments and institutions for groundwater management" in Giordano and Villholth (eds.), *The Agricultural Groundwater Revolution: Opportunities and Threats to Development*, CABI, Oxford.

Kuwayama, Y. and N. Brozović (2012), "Analytical hydrologic models and the design of policy instruments for groundwater-quality management", *Hydrogeology Journal*, Vol. 20(5), pp. 957–972.

Le Kama, A.A. and A. Tomini (2013), "Water Conservation Versus Soil Salinity Control", *Environmental Modeling & Assessment*, Vol. 18 (6), pp. 647-660.

Levy, M. and J. Christian-Smith (2011), "Groundwater Management in the Pajaro Valley", Pacific Institute, Oakland, CA.

Llamas, M. R., and A. Garrido (2007), "Lessons from intensive groundwater use in Spain: Economic and social benefits and conflicts", in Giordano and Villholth (eds.), *The Agricultural Groundwater Revolution: Opportunities and Threats to Development*, CABI, Oxford.

Maliva, R.G. (2014), "Groundwater banking: opportunities and management challenges", *Water Policy*, Vol. 16(1).

Margat, J. and J. van der Gun (2013), *Groundwater around the World: A Geographic Synopsis,* CRC Press/Balkema, Taylor and Francis, London.

Mitsubishi Research Institute, Inc. (2001), Evaluation of Multifunctionality associated with Agriculture and Forestryrelated with the Global Environment and Human Life (in Japanese), Mitsubishi Research Institute Inc., Tokyo.

Napoli, R. and S. Vanino (2011), "Valutazione del rischio di salinizzazione dei suoli e di intrusione marina nelle aree costiere delle regioni Meridionali in relazione agli usi irrigui", Istituto Nazionale di Economia Agraria (INEA), Rome.

OECD (2015), *Water in cities: ensuring sustainable futures*, OECD Studies on Water, OECD Publishing, Paris, http://dx.doi.org/10.1787/9789264230149-en.

OECD (2014a), "Global irrigation water demand projections to 2050: An analysis of convergences and divergences" [ENV/EPOC/WPBWE(2012)2/FINAL], OECD, Paris, http://www.oecd.org/officialdocuments/publicdisplaydocumentpdf/?cote=ENV/EPOC/WPBWE(2012)2/FINAL&docLanguage=En.

OECD (2014b), "Public goods and externalities: Agri-environmental policy measures in Japan" [COM/TAD/CA/ENV/EPOC(2013)56/FINAL], OECD, Paris, http://www.oecd.org/officialdocuments/publicdisplaydocumentpdf/?cote=COM/TAD/CA/ENV/EPOC(2013)56/FINAL&docLanguage=En.

OECD (2013) "Costs and distributional impacts of inaction", in OECD, *Water Security for Better Lives*, OECD Publishing, Paris. DOI: http://dx.doi.org/10.1787/9789264202405-10-en.

OECD (2010), *Sustainable Management of Water Resources in Agriculture*, OECD Studies on Water, OECD Publishing, Pari,. http://dx.doi.org/10.1787/9789264083578-en.

Peck, J.C. (2007), "Groundwater management in the High Plains Aquifer in the USA: Legal problems and innovations", in Giordano and Villholth (eds.), *The Agricultural Groundwater Revolution: Opportunities and Threats to Development*, CABI, Oxford.

Pfeiffer, L. and C.-Y. C. Lin, (2012), "Groundwater pumping and spatial externalities in agriculture", *Journal of Environmental Economics and Management*, Vol. 64(1), pp. 16–30.

Rubio, S.J., and B. Casino (2003), "Strategic Behavior and Efficiency in the Common Property Extraction of Groundwater", *Environmental and Resource Economics*, Vol. 26, pp. 73-87.

Saak, A.E. and J.M. Peterson (2007), "Groundwater use under incomplete information", *Journal of Environmental Economics and Management*, Vol. 54(2), pp. 214–228.

Scanlon, B. R. et al. (2012), "Groundwater depletion and sustainability of irrigation in the US High Plains and Central Valley" *Proceedings of the National Academy of Sciences of the United States of America*, Vol. 109(24), pp. 9320–5. doi:10.1073/pnas.1200311109.

Schoengold, K. and D. Zilberman (2007), "The economics of water, irrigation, and development", *Handbook of Agricultural Economics*, Vol. 3(06).

Scott, C., S. Dall'erba and R.D. Caravantes (2010), "Groundwater Rights in Mexican Agriculture: Spatial Distribution and Demographic Determinants", *The Professional Geographer*, Vol. 62(1), pp. 1–15.

Scott, C.A. and T. Shah (2004), "Groundwater overdraft reduction through agricultural energy policy: insights from India and Mexico", *International Journal of Water Resources Development*, Vol. 20(2), pp. 149–164.

Séguin, J.J., E. Gomez, E. and A. Wuilleumier (2009), "Les indicateurs piézométriques: un outil dans la gestion des hydrosystèmes. Orientations méthodologiques", BRGM, Nancy.

Shah, T, J. Burke, K. Villholth and M. Angelica (2007), "Groundwater: a global assessment of scale and significance", in D. Molden (Ed.), *Water for Food, Water for Life: A Comprehensive Assessment of Water Management* (pp. 395–423), International Water Management Institute, Sri Lanka.

Shah, T. (2008), *Taming the anarchy: groundwater governance in South Asia, Resources for the Future*, Washington, DC, and International Water Management Institute, Colombo, Sri Lanka.

Siebert, S. et al. (2010), "Groundwater use for irrigation – a global inventory", *Hydrology and Earth System Sciences*, Vol. 14(10), pp. 1863–1880.

Sneed, M., J. Brandt and M. Solt (2013), "Land Subsidence Along the Delta-Mendota Canal in the Northern Part of the San Joaquin Valley, California, 2003-10", *Scientific Investigations Report*, 2013–5142, United States Department of Interior and United States Geological Survey, Washington, DC, USA, http://pubs.usgs.gov/sir/2013/5142/.

Snyder, J. (1955), "Ground Water in California: The Experience of Antelope Valley", Giannini Foundation Ground Water Series No. 2. http://ageconsearch.umn.edu/bitstream/147028/2/Snyder%20-%20Antelope%20Valley%20-%201955%20ocr%20version.pdf.

Sophocleous, M. (2012), "Conserving and extending the useful life of the largest aquifer in North America: the future of the High Plains/Ogallala aquifer", *Ground water*, Vol. 50(6), pp. 831–9.

Taniguchi, M., W.C. Burnett, and G.D. Ness (2008), "Integrated research on subsurface environments in Asian urban areas", *Science of the Total Environment*, Vol. 404, pp. 377-392.

Taylor, R.G. et al. (2012), "Ground water and climate change", *Nature Climate Change*, Vol. 3, pp. 322-329.

Teatini, P., M. Ferronato and G. Gambolati (2006), "Groundwater pumping and land subsidence in the Emilia-Romagna coastland, Italy: Modeling the past occurrence and the future trend", *Water Resources Research*, Vol. 42: W014066.

UNESCO (United Nations Educational, Scientific, and Cultural Organization) (1984) *Guidebook to studies of land subsidence due to ground-water withdrawal*, edited by J.F. Poland, UNESCO – IHP, Paris.

van der Gun, J. (2012), "Groundwater and Global Change: Trends, Opportunities and Challenges", United Nations World Water Assessment Programme, UNESCO, Paris. http://www.unesco.org/new/fileadmin/MULTIMEDIA/HQ/SC/pdf/Groundwater%20and%20Global%20Change.pdf.

Vrba, J., and A. Lipponen (eds.) (2007), *Groundwater resources sustainability indicators*, IHP-VI, Series on Groundwater No 14, UNESCO-IHP, Paris, France.

WHYMAP (World-wide Hydrogeological Mapping and Assessment Programme) (2004a), "Groundwater Resources of the World 1:50000", Bundesanstalt für Geowissenschaften und Rohstoffe (BGR), Hannover, Germany. http://www.whymap.org/whymap/EN/Home/whymap_node.html.

WHYMAP (2004b), "Statistics", Bundesanstalt für Geowissenschaften und Rohstoffe (BGR), Hanover, http://www.whymap.org/whymap/EN/Statistics/statistics_node_en.html.

Wijnen et al. (2012), "Managing the Invisible: Understanding and Improving Groundwater Governance", The Water Partnership Programme, World Bank, Washington, DC.

Winter, T.C. et al. (1998). "Ground Water and Surface Water", U.S. Geological Survey Circular 1139, USGS, Denver, CO.

Annex 2.A1

Existing typologies on groundwater and irrigation systems

Table 2.A1.1. IGRAC groundwater regions in OECD countries and their characteristics

IGRAC region	OECD countries	Class	Main geology	Climate	Groundwater resources
1. Western Mountain belt of North and Central America	Canada, United States, Mexico	High relief folded mountains region	Basis of sedimentary and metamorphic rocks and volcanic rocks	From permafrost to oceanic and arid	Variable groundwater resources. Fluvial aquifer: (California Central Valley) Coastal aquifers. (Baja California)
2. Central plains of North and Central America	Canada, United States, Mexico	Sedimentary basin region	Thick layers of sedimentary rocks	Primarily dry	Rich resources, major aquifers: Ogallala, Northern Great Plains
3. Canadian Shield	Canada	Basement region	Crystalline rocks and a few sedimentary basins	Snow, permafrost	Limited resources
4. Appalachian highlands	Canada, United States	High relief folded mountains region	Metamorphic rocks with sedimentary basins	Humid	Variable resources mostly in carbonate rocks and sandstone aquifers, plus alluvial shallow aquifers
5. Caribbean islands and coastal plains of North and Central America	United States, Mexico	Sedimentary basin region	Alluvial and marine sedimentary plains, superimposed by volcanic rock (Caribbean)	Humid	Abundant resources in Alluvial sedimentary basins, largely karstic and some carbonate and volcanic aquifers
6. Andean Belt	Chile	High relief folded mountains region	Metamorphic, granitic, volcanic and sedimentary	Variable from humid to dry	Variable. Coastal sedimentary and volcanic aquifers.
10. Baltic and Celtic Shields	Estonia, Finland, Sweden, Norway, Iceland, Ireland, United Kingdom, France	Basement region	Mainly Crystalline rocks, Sedimentary (EST, IRL), volcanic (ISL).	Medium to highly humid	Limited groundwater resources. Local karstic and volcanic aquifers.
11. Lowlands of Europe	United Kingdom, France, Belgium, Luxemburg, Netherlands, Denmark, Germany, Poland	Sedimentary basin region	Thick sedimentary plains	Medium humid	Abundant resources. Major aquifer (Paris Basin), limestone aquifer (Chalk aquifer in UK), sandstone aquifers.
12. Mountains of Central and Southern Europe	Portugal, Spain, France, Germany, Switzerland, Austria, Italy, Czech Republic, Slovak Republic, Slovenia, Hungary, Greece.	High relief folded mountains region	Crystalline, volcanic and sedimentary structures	Dry to humid (Alps)	Variable resources, significant sedimentary basins (Po valley, Hungarian Plains)

Table 2.A1.1. IGRAC groundwater regions in OECD countries and their characteristics (*cont.*)

IGRAC region	OECD countries	Class	Main geology	Climate	Groundwater resources
23. North-western Pacific margin	Japan	Volcanic region	Sedimentary and volcanic rocks	Variable from dry to humid	Variable. Productive volcanic and sedimentary aquifers (Tokyo)
24. Mountain belt of Central and Eastern Asia	Korea	High relief folded mountains region	Crystalline, sedimentary rocks	Humid in coastal areas	Variable resources: Karstified carbonate aquifers
26. Mountain belt of West Asia	Turkey	High relief folded mountains region	Crystalline, volcanic and sedimentary rocks	Dry	Variable. Significant resources in karstified limestone (Midyat aquifer in Turkey)
31. Levant and Arabian platform	Israel	Sedimentary basin region	Sedimentary valleys	Arid	Abundant but not renewable limestone complexes in the Mediterranean
34. Western Australia	Australia	Basement region	Crystalline rock, sandstone, karstified limestone and alluvial sediments	Arid to tropical humid (North)	Limited to moderate resources. Fissured sandstone (Canning aquifer) and limestone
35. Eastern Australia	Australia	Sedimentary basin region	Sedimentary alluvial formations	Arid to semi-arid more humid towards the coast	Moderate to high, major sandstone aquifer (Great Artesian Basin), Shallow alluvial sedimentary aquifers.
36. Islands of the Pacific	New Zealand	Volcanic region	Crystalline and sedimentary rocks (New Zealand)	Humid	Variable resources, some volcanic aquifer, and sedimentary (alluvial, marine) regions have significant resources.

Source: IGRAC (2004) and Margat and van der Gun (2013).

Table 2.A1.2. Proposed typology of groundwater economies by Shah et al. (2007)

	Arid agricultural systems	Industrial agricultural systems	Smallholder farming systems	Groundwater-supported extensive pastoralism
Countries	Algeria, Egypt, Iran, Iraq, Libya, Morocco, Tunisia, Turkey	Australia, Brazil, Cuba, Italy, Mexico, South Africa, Spain, United States	Afghanistan, Bangladesh, People's Republic of China, India, Nepal, Pakistan	Botswana, Burkina Faso, Chad, Ethiopia, Ghana, Kenya, Malawi, Mali, Namibia, Niger, Nigeria, Senegal, South Africa, Tanzania, Zambia
Groundwater-irrigated areas	Less than 6 million hectares	6-70 million hectares	71-500 million hectares	More than 500 million hectares supported by boreholes for stock watering
Climate	Arid	Semiarid	Semiarid to humid, monsoon climate	Arid to semiarid areas
Aggregate national water resources	Very small	Good to very good	Good to moderate	Mixed rain fed livestock and cropping systems
Population pressure on agriculture	Low to medium	Low to very low	High to very high	Low population density but pressure on grazing areas is high
Share of total land area under cultivation[1]	1-5%	10-50%	40-60%	5-8%
Share of cultivated areas under irrigation[1]	30-90%	2-15%	40-70%	<5%
Share of irrigated area under groundwater irrigation[1]	40-90%	5-20%	10-60%	<1%
Share of total geographic area under groundwater irrigation[1]	0.12-4.0%	0.001-1.5%	1.6-25.0%	<0.001% but groundwater supported grazing areas about 17% of total
Organization of agriculture	Small to medium size farms under market based agriculture	Medium size to large scale farms under industrial, export-oriented farming	Very small landholdings, subsistence-oriented, mixed peasant farming systems	Small-scale pastoralists, often seasonally connected with small-scale agriculturalists
Driver of groundwater irrigation	Lack of alternative irrigation or livelihood	Highly profitable market-based farming	Need to absorb surplus labour in farming through land-augmenting technologies	Stock watering
Significance of groundwater irrigation to national economy	Low (<2-3% GDP)	Low (<0.5% GDP)	Moderate (5-20% GDP)	Moderate (5-20% GDP)
Significance of groundwater irrigation economy to welfare of national population	Low to moderate	Low to very low	Very high (40-50% of rural population and 40-80% of food production involve groundwater irrigation)	Low in terms of numbers of pastoralists involved, sometimes moderate in terms of national food supply
Significance of groundwater irrigation for poverty reduction	Moderate	Very low	Very high	Groundwater central to pastoral livelihood systems, but limited scope for using more groundwater for poverty reduction
Gross value of output supported by groundwater irrigation	USD 6-8 billion	USD 100-120 billion	USD 100-110 billion	USD 2-3 billion

Note: 1. Ranges of estimates provided for these rows correspond to the categorisation by Shah et al. (2007); they do not necessarily reflect actual estimates (but rather the approximations they made).

Source: Shah et al. (2007).

Table 2.A1.3. The GW-MATE typology of groundwater systems

Overall typology of groundwater body	Sub-divisions by type of situation or process involved
(1) At risk of extensive quasi-irreversible aquifer degradation and subject to potential conflict amongst users	(a) Under intensive exploitation
	(b) Vulnerable to widespread pollution from land surface
	(c) Undergoing depletion of non-renewable storage reserves
(2) Subject to potential conflict amongst users but not at risk of quasi-irreversible aquifer degradation	(a) With growing large-scale abstraction
	(b) Vulnerable to point-source pollution
	(c) With shared international/interstate resources
(3) Insufficient (or inadequate use of) scientific knowledge to guide development policy and process	(a) But potential to improve rural welfare and livelihoods
	(b) With presence of natural quality problems
	(c) But scope for large-scale planned conjunctive use

Source: Foster et al. (2009).

Table 2.A4. Proposed typology of groundwater and surface water resources use in irrigation depending on climatic conditions

	Favourable conditions for groundwater withdrawals (high transmissivity and storage volume)	**Unfavourable conditions for groundwater withdrawals** (low transmissivity and storage volume)
Unfavourable climatic conditions (low groundwater recharge)	Irrigation using recharge from surface water and groundwater (if surface water generated in areas with favourable climate is available) or irrigation using non-renewable groundwater from deep wells	Surface water irrigation (from canals, rivers or reservoirs) using runoff generated in areas with favourable climatic conditions
Favourable climatic conditions (high groundwater recharge)	Irrigation using mainly renewable groundwater from springs and wells	Surface water irrigation (from canals, rivers or reservoirs)

Source: Siebert et al. (2010).

Chapter 3

What policy instruments help to manage agricultural groundwater use sustainably?

This chapter provides a normative analysis of agricultural groundwater management policies. Through an economic lens, it reviews the rationale for groundwater public policies and management, and discusses the advantages and drawbacks of the main instruments used to manage groundwater in agriculture.

The statistical data for Israel are supplied by and under the responsibility of the relevant Israeli authorities. The use of such data by the OECD is without prejudice to the status of the Golan Heights, East Jerusalem and Israeli settlements in the West Bank under the terms of international law.

Key messages

Public policy is needed to address externalities generated by intensive groundwater use and cope with long-term depletion. In other cases, such as temporary depletion or the management of renewable groundwater as complements to surface water, the specificity of local groundwater systems (Chapter 2) and associated constraints should determine when policy intervention is needed.

Among the range of instruments available to manage these challenges, demand-side instruments should be prioritised as they tackle more long-term incentives; supply-side instruments should be used in a complementary manner There are multiple policy instruments designed to either reduce demand or increase supply of water for irrigators, directly or indirectly. Of these, demand-side instruments, such as quantitative reduction, well regulations or pricing, act on users' incentives, while supply-side instruments, such as recharge, or surface water storage, only relieve a constraint without affecting the underlying incentives that determine production systems.

In relation to groundwater management, no single policy instrument alone can respond to the different settings. Even within a single region, the welfare rankings of alternate policies may vary depending on desired environmental and hydrological goals. Effective policy analysis requires an understanding of the economic drivers underlying decision-making related to groundwater use, the potential social welfare and environmental impacts of alternative policies, and the implications for longer-term aquifer management.

Policy makers should focus on comprehensive adaptive management. Improving information collection and monitoring, lowering transaction costs, and designing locally customised options within existing regulatory frameworks. Local involvement may help to enhance the effectiveness and implementation of policy instruments.

Different policy approaches have a role to play within a management framework and should be adapted to local circumstances.

- On the demand side, regulatory measures generally face information constraints that prevent targeted and effective regulation; economic schemes can lead to efficient outcomes but may be associated with high transaction costs; and collective management schemes are intrinsically adapted to local constraints but depend on participation and commitment of users.

- Supply-side approaches may provide complementary approaches to resolve binding constraints but can be costly and therefore require significant financing and investment.

- Each of these instruments also requires data and information collection, effective monitoring of users, and enforcement.

Chapter 4 will review current agricultural groundwater policies in OECD countries, while Chapter 5 will discuss how they perform in light of the conclusions above.

Looking for efficient and effective management solutions

Groundwater is an important resource for agricultural water users, representing about a quarter of freshwater withdrawals worldwide and largely contributing to irrigation in several OECD countries (Chapter 1). Its use is often unmonitored and unregulated. However, as reported in Chapter 2, there may be negative consequences of groundwater overdraft in specific groundwater systems, including on neighbouring wells, on adjacent stream flows, and on the future availability of water supplies for growing populations (Chapter 2).

Addressing these challenges in an economically efficient way is a key question for increasingly concerned agricultural and environmental policy makers. This chapter discusses the economics of agricultural groundwater management. Building on lessons from the literature and on a simplified economic model, it reviews and analyses the main policy instruments with an emphasis on addressing long-term groundwater depletion and externalities. It does not aim to provide a comprehensive review of groundwater policies and management practices in OECD countries; that is the object of Chapter 4. Instead, its objective is to help understand and evaluate public policies surrounding agricultural groundwater management.[1]

The first section of this chapter discusses the scope of groundwater management. The second section presents an overview of management approaches and the third section employs a simple economic model to outline what factors may matter. The fourth and fifth sections analyse the main policy instruments used for groundwater management on the demand and supply sides.

Scope for public action: Managing long-term depletion and externalities

The first core question that precedes policy analysis is the justification for intervention. Public economic theory supports actions in the presence of market imperfections and market failures. But the definition of public versus private goods is not always clear in the case of water (OECD, 2015), and perhaps even more complex for the specific case of groundwater (Mechlem, 2012).

Groundwater is often considered a common pool resource (Foster et al., 2009; Lopez-Gunn et al., 2012a), i.e. defined by the presence of costly exclusion and subtractability of units. Each unit that is extracted by a user is not available for others (Schlager, 2007). This definition, however, can be deceiving. Strict common pool and private property resources are end-members along a spectrum, and most aquifers will fall somewhere along the continuum. As such, attributing a common pool resource status to groundwater is often not applicable as it depends on the nature of the aquifer (Brozovic et al., 2006). An aquifer with high storativity[2] and low transmissivity is closer to a private property than common pool resource (Huang et al., 2012). As a result, gains from management interventions may be less than expected and quite variable (Brozovic et al., 2006). Furthermore, the degree of connection with surface water systems can also affect whether groundwater acts as a private or common pool resource.

Common pool resources face provision and appropriation problems, defined respectively as the way to ensure that the resource is maintained and preserved and the challenges associated with resource allocation. There are different institutional responses to these problems, ranging from co-operative management institutions to nonco-operative or exogenous regulatory institutions (Madani and Dinar, 2013). While provision problems will most often require government intervention, appropriation challenges can in some case be resolved by users themselves (Schlager, 2007).

When does management result in welfare gains? In the case of groundwater management as a renewable resource, the literature provides no general answer. A number of economic articles have studied the question (Koundouri, 2004; Roumasset and Wada, 2013), largely focusing on the issue of groundwater abstraction or as a case of renewable resource management, relying on relatively simplified settings, assuming no externality, relatively well-defined property rights, and relatively large aquifers (high storativity, low transmissivity) (e.g. see Gisser and Sanchez, 1980). The conclusion of their work is that that the benefits of management or groundwater irrigation will not always be significant, and that they depend highly on economic, hydrologic, and agronomic parameters (Koundouri, 2004).

However, just as in other areas of environmental policies, agriculture groundwater management is unambiguously called for when facing negative externalities that are not accounted for by users. There has been significant discussion in the literature on possible criteria to define when groundwater pumping is detrimental (see Chapter 2, and Llamas and Garrido, 2007). In particular, fighting against stock depletion is rarely found to be appropriate (Giordano, 2009; GWP, 2012). But addressing pumping-induced externalities is a commonly agreed objective in the literature (e.g. Garduño and Foster, 2010; Llamas, 2004; Llamas and Martinez-Santos, 2005; OECD, 2013a). As noted in Chapter 2, there are multiple types of externalities, with different degrees of damages, and only applicable under specific groundwater systems. But even the most common cases of well yield reduction and increased pumping costs may call for management, especially if their aggravation can lead to more important consequences.

At the same time, long-term groundwater depletion (of relatively large-scale aquifers) will generally call for public policy intervention. In such cases, common pool resource properties may become more prevalent as the average resource disappears. Quality concerns and externalities that affect ecosystems are likely to increase as the level goes down. And the provision problem that depletion creates may call for long-term planning and management to avoid future appropriation problems.

These rationales reflect actual practices. Over the last decade, there have been two broad categories of concerns underlying policy change in groundwater management, especially in OECD countries. First, concerns about the physical decline of aquifer systems include changes in both groundwater quantity and groundwater quality available (including salinity), as well as the potential for irreversible land subsidence (e.g. Konikow, 2013). Second, interactions between groundwater and surface water systems have also been a major driver for changes in groundwater policy. For example, concerns over stream depletion have led to the introduction of regulations on groundwater use in a number of transboundary river basins in the United States, including the Pecos River (between Texas and New Mexico), Arkansas River (Kansas and Colorado), and the Republican River (Kansas, Nebraska, and Colorado), as well as in other countries such as the Guadalquivir Basin in Spain. The adverse effects of stream depletion on instream habitat and endangered species have also led to regulatory action, for instance in multiple US states.

In contrast, interference between adjacent pumping wells, and the potential for strategic behaviour by producers in response to such interference, while an important focus of economic research over the last thirty years, has with few exceptions not been an important driver of binding policy change in groundwater management. One reason for this may be that – at least as large-scale irrigated agriculture is practiced in many OECD member countries – well spacing is large enough that the external effects associated with pumping imposed on neighbouring well owners is relatively small compared to drawdown induced by a well's own pumping (Brozović et al., 2010). Indeed, well spacing regulations are one kind of policy that is locally tailored precisely to reduce the potential for well interference. Moreover, when interviewed, producer concerns generally do not focus on their neighbours' potential strategic behaviour (Dixon 1989). Thus, this chapter will not focus on strategic behaviour as either a key motivation of agricultural producers or as a driver of policy change.[3]

Choice of policy instruments: A wide range of options

Managing groundwater require a good understanding of flows and the effects directly associated with pumping to identify what leverage a policy can have. Box 3.1 provides a simple conceptual model of an unconfined aquifer with linked surface water systems to illustrate key aspects and dynamics of the coupled "natural-human system". This simplified model shows: a) that there are multiple flows in groundwater systems; b) that different wells can have different effects, depending on their relative location and depth; and c) that crop irrigation practices may also matter.

Box 3.1. A simple model of human-natural interactions in an aquifer

Figure 3.1. A cross-sectional diagram of interactions in a linked surface water-groundwater system

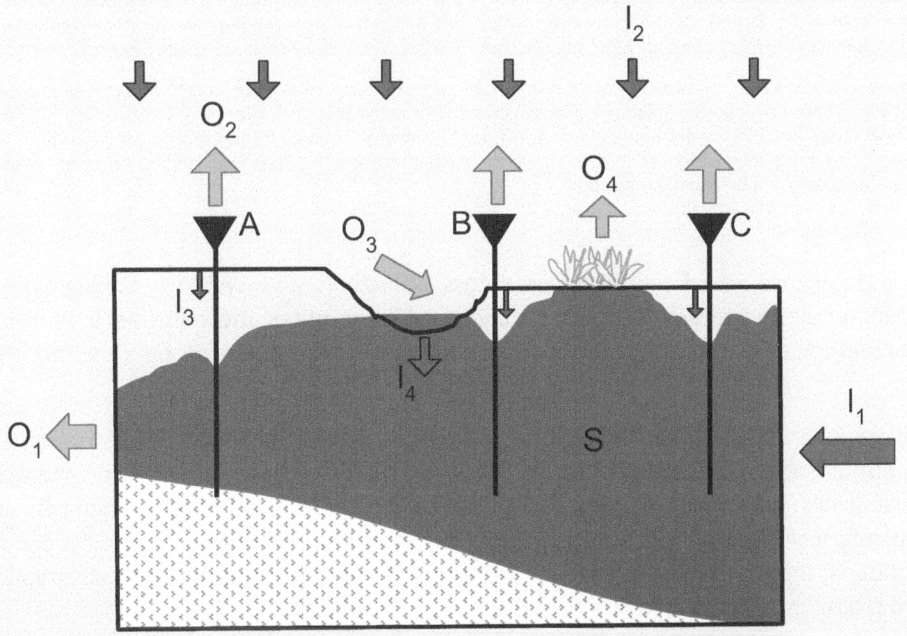

Note: A, B and C are three wells. S is the aquifer and the cross-hatching area under S is a saline lens. Unit pumping costs will be largest for well A as the depth to groundwater is the largest. Moreover, the saturated thickness of the aquifer underlying well A is also less than that for wells B and C, suggesting that well A will have the lowest well yield (instantaneous application rate of water).

Figure 3.1 represents a conceptual cross-section of a linked surface water-groundwater system. In the figure, there are three wells (denoted **A**, **B**, and **C**) that are used to pump groundwater for agricultural production. There are a number of inputs to the regional aquifer system. Regional groundwater inflow into the management area is denoted $\mathbf{I_1}$, and the portion of precipitation that recharges the aquifer (percolation from precipitation) is $\mathbf{I_2}$. Return flow from irrigation is denoted $\mathbf{I_3}$ and $\mathbf{I_4}$ is recharge from the stream system interacting with the aquifer. In terms of outputs, $\mathbf{O_1}$ is the regional groundwater outflow, $\mathbf{O_2}$ is crop evapotranspiration, $\mathbf{O_3}$ is surface and subsurface flow into the stream system, and $\mathbf{O_4}$ is water for native vegetation evapotranspiration and needed to support groundwater-dependent ecosystems. A mass balance for the aquifer over any given time period relates the sum of inputs ($\mathbf{I_1} - \mathbf{I_4}$), the sum of outputs ($\mathbf{O_1} - \mathbf{O_4}$), and the change in available water stored in the aquifer ($\Delta\mathbf{S}$):

$$\sum_{n=1}^{4} \mathbf{I_n} - \sum_{n=1}^{4} \mathbf{O_n} = \Delta\mathbf{S}$$

This mass balance equation does not specify the length of time represented, so it is equally valid for short-run and long-run analyses. This conceptual figure allows us to analyse both the regional and local impacts of agricultural groundwater use on the surface water-groundwater system. The mass balance makes clear that for a steady state of the aquifer — defined as no change in available water stored, or $\Delta S = 0$ — to exist, the inputs and outputs to the system must balance. If the total outputs from the surface water-groundwater system exceed the total inputs, the change in storage will be negative and the aquifer will be depleted. If total inputs exceed total outputs, the aquifer will recharge.

Box 3.1. A simple model of human-natural interactions in an aquifer *(cont.)*

From a regional perspective, if subsurface inflow into the system (I_1) or precipitation (I_2) decrease, then even if groundwater pumping (O_2) does not increase, there will be either a decrease in system outflow or a decrease in the contribution of groundwater to the stream's baseflow (O_3) or groundwater-dependent ecosystems (O_4). Reductions in streamflow may lead to impacts on both instream habitat and to downstream or transboundary conflict over shared surface water allocations. If agricultural producers increase the efficiency of their irrigation systems while keeping pumping the same, recharge I_3 will decrease, and once again there will be a reduction in outputs, storage, or a combination of the two. Importantly, it is expected in a drought year that all inputs will decrease, but that both crop and vegetation water demands will increase, leading to reductions in streamflow for rivers that are hydrologically connected to the aquifer, as well as reduced water availability for other groundwater-dependent ecosystems.

In addition to regional impacts on the groundwater system, there may also be localised impacts from groundwater pumping. These spatial externalities may include stream depletion and local water table lowering. For example, ongoing pumping by well **B** will lead to stream depletion and reductions in streamflow. As well, as **B** is closer to the river than wells **A** and **C**, the impact of pumping will be larger for every unit of water pumped by well **B** than the other wells. On the other hand, both wells **B** and **C** can potentially affect the groundwater-dependent ecosystem (responsible for O_4 in Figure 3.1) if their respective cones of depression lower the water table that the ecosystem depends on.

If saline water underlies the freshwater aquifer then ongoing aquifer depletion may lead to reductions in water quality at a pumping well, compromising the ability to irrigate crops. For example, in Figure 3.1 there is a lens of saline water (denoted with cross-hatching in the figure) underlying the freshwater aquifer. As the amount of storage in the freshwater aquifer is reduced, the saline water will intrude upwards. As it intersects the screened portion of a well — as observed for well **A** — water quality will be rapidly reduced.

Even if there are specific differences across objectives and systems, coping with long term groundwater depletion and/or associated externalities requires mechanisms to control pumping and/or increasing access to alternative water sources. Two types of policies can therefore be employed:

- Policies operating on the *demand side* to reduce water consumption. Instruments can focus on authorisation to use wells (i.e. operating on the extensive margin) or on the actual use of these wells (intensive margin), using direct or indirect approaches. Specific agricultural controls, including land use, crop choice, irrigation type, or energy contracts, can be used as additional indirect levers to control groundwater use. Agriculture and conservation policies and related practices can also affect the use of groundwater.

- Policies operating on the *supply side* to augment water availability for irrigation and other uses. Such approaches will try to increase groundwater storage and/or using alternative sources of water (e.g. surface water, treated waste water). Increasing access to surface water for irrigation, via infrastructure investments, and use or storage of recycled wastewater are among the recently encouraged approaches, while desalinization and groundwater banking are still not widely used in agriculture due to costs of entry they entail.

Table 3.1 provides a rapid overview of approaches and instruments. For relative consistency and tractability, three main types of management approaches are distinguished: regulatory, economic, and collective management approaches, acknowledging that there are a number of other proposed classifications in the literature.[4] In practice, there may be some overlap between types; for example, strictly enforced regulations often underlie economic schemes, and the latter rely on a sufficient regulatory basis. Furthermore, instruments are often combined rather than applied individually. Each of the cells provides selected examples of public instruments that could affect groundwater management in agriculture. It, however, excludes private approaches supported by legal instruments, such as adjudication or litigation, that do not fit into any of the above-discussed categories.

Table 3.1's vast number of potential instruments creates a space for arbitrage and guidance. As noted in Chapter 2, just as groundwater systems are diverse and associated with specific challenges, not all options will be useful in all contexts. The economics of agricultural groundwater management provide a useful basis for decision. The next section proposes elements of a model to help identify factors of differentiation for choice. The following sections then build on this framework to evaluate some of the key management options.

Table 3.1. Main types of instruments used to manage groundwater use in agriculture

		Regulatory approaches	Economic instruments	Collective management approaches
Demand-side approaches	*Extensive margin (wells)*	Permit requirement for wells	Well taxes	Shared investment and decision to drill wells
	Intensive margin (use)	*Direct*: Entitlements, mandatory monitoring, quotas, minimum efficiency	*Direct*: Groundwater pricing, markets for groundwater entitlements; water efficiency schemes	*Direct:* Self-designed voluntary schemes to reduce groundwater pumping
		Indirect: Regulations on land, conservation and energy use[1]	*Indirect:* Agriculture support programs, payment for environmental practices and voluntary conservation schemes, energy and land fiscal and market measures[1]	
Supply-side approaches	*Additional supply for storing*	Regulatory storage objectives with fines and penalties	Cost sharing programs, loans, subsidised groundwater banking	Cost-shared infrastructure building, aquifer storage and recovery programs
	Additional supply for use	Surface water reallocation	Financing infrastructure (dams, treatment, desalinization).	Collective management plans, recycled water use for irrigation

1. Indirect instruments are presented here as intentional approaches but can also be introduced for other purposes and unintentionally affect groundwater use.

What factors count in the choice of instruments?

A simplified economic model to develop key intuitions

For the purposes of this chapter, the main features and findings of a simple model of agricultural groundwater use is presented in order to develop key intuition about alternate groundwater management policies. The model is presented in details and full notation in Annex 3.A. It is based on Brozović and Young (2014) and focuses primarily on the use of demand-side instruments to alter farmer use in a groundwater basin to respond to an external constraint (here, overall or stream depletion, but could be adapted to other). A number of caveats apply to this model, which are generally common in the literature:

- It focuses on aggregate water applied during an entire growing season for analysis.
- It abstracts from complex aquifer dynamics and strategic behaviour (e.g. Saak and Peterson 2007, Athanassoglou et al., 2012).[5]
- Groundwater used is assumed to be measurable.
- Individual well benefit functions are assumed to be independent, so that pumping at one well will not affect pumping at other wells.[6]
- Users are assumed to be risk-neutral.

Incorporating more realistic spatial dynamics provides qualitatively similar results with more complex optimality conditions (e.g. Brozović et al., 2010; Kuwayama and Brozović, 2013). The strategic aspect provides some interesting theoretical outcomes, but may not represent the situation for a large majority of aquifers.

The core of the micro-economic model is founded on the optimisation of pumping behaviour at each well. It is, therefore, based on a number of wells that represents users; each has a benefit function for pumping that depends on well-specific parameters (soil type, irrigation technology, field-level crop management choices) and incorporates a crop production function. For example, all else being equal, a producer's costs of production will increase as depth to groundwater increases or well yields decrease. A field with sandy soil will require that more water to be applied to the crop in order to give the same yield as a field with a finer-grained soil. The benefit function also incorporates parameters that affect all users, such as input and output prices and weather.[7] For example, in a dry year, it is expected that the benefit from applying any amount of water to a crop will be larger than in a wet year.

In this setting, the overall net benefit function is equal to the sum of individual benefits. The general groundwater management problem is to maximise the sum of economic benefits of agricultural groundwater pumping for each well, subject to constraints related to the hydrologic impacts of agricultural pumping.[8]

Adjusting agricultural groundwater use to limit overall aquifer depletion

If the main management concern is aquifer depletion, then the goal of regulation is to reduce pumping in each year below a given amount that is specified by a hydrologic model, e.g. to the amount needed to maintain baseflow or preserve water tables needed for groundwater-dependent ecosystems. From a management perspective, it may be convenient for the total amount pumped to be the same each year, but this is not necessary.[9]

The problem is not a first-best social optimisation of the coupled natural-human system (Box 3.1). First-best optimisation would require explicit valuation of all possible environmental services associated with the groundwater resource, both now and in the future. Instead, society decides on the level of hydrologic services that is desirable and the economic problem is to achieve that level using a solution that maximises benefits to water users. This process involves an implicit valuation of environmental services, but not an explicit one. Such a decision-making framework corresponds more closely to the groundwater management problem observed in the real world than a first-best optimisation approach.

In the simplest case, all pumping can be assumed to have an equal impact on the aquifer in terms of saturated thickness. Examples of such aquifers include single-cell aquifers such as that modelled by Gisser and Sanchez (1980) or more complex aquifers where the only issue of concern to policy makers is the overall amount of water stored in the system. The management problem may then be stated as a maximisation of the benefits subject to an aggregate pumping constraint.

As a result, it can be shown that an optimal allocation of water across all constrained groundwater users is one that equates their marginal benefits at each point in time. In particular, in the absence of a pumping constraint, this model would predict that each producer would pump until the value of the marginal product of water (or marginal benefit of pumping) was equal to zero. As the pumping constraint becomes more binding, the value of the marginal product of water will increase. The optimal choice of marginal benefit depends on the degree to which pumping must be constrained to meet the aggregate pumping target: the lower the desired value of the target, the higher the marginal benefit needed to achieve it, as each producer must be constrained more.

If the hydrologic constraints are more complicated, for example if the aquifer properties vary across space, or if it is desirable to limit pumping further in some localised zones, such as around drinking water supplies, then the general hydrologic constraint may be altered so that it incorporates the required spatial heterogeneity. So long as the impacts of pumping are effectively independent, the optimal conditions will be obtained when *the ratio of marginal benefit to the marginal impact of the hydrological constraint will be equalised across all users*. For a well that has a relatively higher impact on the aquifer, the marginal benefit at the optimal allocation will likewise be higher in order to satisfy the optimality condition, i.e. less pumping will be allowed.

Dynamic management to control aquifer depletion or stream depletion

In this case, the assumed goal of groundwater management is to choose a set of pumping paths to address issues of either aquifer depletion or stream depletion. The general management problem is then stated as the maximisation of benefits for each user over each increment of time, subject to the defined hydrological constraints and characteristics for each users and periods at a specific time. The key variables defined above are allowed to vary over time. As climate is highly variable both during a growing season and between seasons, it follows that the economic value to agricultural producers of their ability to apply water to crops will also vary enormously through time.[10]

Externalities due to stream depletion caused by groundwater pumping may lead to the need for management if there are downstream or transboundary legal obligations related to surface water or if there instream impacts on habitat. In the case of downstream or transboundary surface water obligations, the intent of regulation is generally to reduce pumping in order to limit cumulative stream depletion over a fixed interval such as a year or multiple years. In the latter case, regulations are intended to maintain minimum streamflow requirements throughout the year.

A key feature of surface water-groundwater interaction is that stream depletion is a spatial and dynamic process and that because groundwater is a diffusional system, it is also subject to lagged effects (Glover and Balmer, 1954; Sophocleous, 2002). Thus, the impact of ongoing pumping on streamflow needs to consider the pumping history rather than just the pumping that is occurring in the current period. The total stream impact from groundwater pumping at any time after the start of pumping is then equal to the sum of lagged impacts occurring at that time from all pumping that occurred at or before this time, accounting for well- specific distances from the stream.

Hydrologic stream response functions can then be used to model the exact relationship between pumping and stream flow. Both analytical and numerically-derived methods are currently in use in implemented regulations to determine stream response to groundwater pumping. Where detailed numerical groundwater models are available, these have been used to determine the impact of pumping on stream depletion. Elsewhere, analytical and graphical methods based on solutions of the groundwater flow problem applied to the case of surface water-groundwater interaction are applied (as shown in Box 3.2).

Hydrologists have derived stream response functions for a variety of different hydrologic settings. In particular, the analytical solution by Glover and Balmer (1954), while being one of the simplest, has been widely applied in a policy context (e.g. Jenkins, 1968; Nebraska DNR, 2007). This equation allows to derive the stream depletion caused by a specific well after a defined pumping period at a constant rate of impacts based on pumping rates, distance from the stream, aquifer coefficients (storage, transmissivity) and a complementary error function (see Annex 3.A for details). The equation can be modified to account for seasonal pumping. Similarly, other more sophisticated versions are available for surface-water groundwater interactions such as partially penetrating wells or streambed clogging (e.g. Hunt, 1999; Hunt, 2012).

Box 3.2. Applying analytical models to address stream depletion: Examples from the United States

Several US states have determined areas where groundwater is hydrologically connected to adjacent rivers and have used this definition for regulatory design. In some cases, entire watersheds are given a designation of connectivity, but in others, a combined spatial and temporal definition is used. For example, the Nebraska Department of Natural Resources (DNR) has implemented the "10/50 rule" (Nebraska DNR, 2007) which defines separate zones, and therefore potential regulations, for wells based on whether or not groundwater pumping over a 50-year period will include at least a 10% contribution from an adjacent stream. In some cases, Nebraska has also applied a 28/40 rule. This rule defines zones based on wells expected to pump at least 28% of their water from an adjacent stream over a 40-year period (Nebraska DNR, 2004). As stream depletion increases with both time and proximity to a stream, all else being equal, the 10/50 rule is more stringent than the 28/40 rule and will cover a large area adjacent to streams where stream depletion is a concern. In Nebraska, numerical methods have been used in the Republican River Basin and the Big Blue River Basin (MODFLOW-based), and in the Platte River Basin (COHYST-based).

In addition to their current use for designing groundwater regulations, analytical methods are used by practitioners for general assessments of stream depletion. In Kansas, analytical methods have been used to determine whether additional groundwater is available for appropriation. For the Lower Republican River Basin and Belleville Formation in Kansas, the Jenkins method (a graphical approach based on the Glover-Balmer equations; Jenkins, 1968) has been used to estimate the cumulative volume of stream depletion that occurs in one year after the day pumping begins for an application to appropriate groundwater to see whether the new appropriation is acceptable (Kansas Department of Agriculture, 2010). The Glover and Balmer method has been employed in Colorado to evaluate the current and projected stream depletion impacts of water pumped and discharged during coalbed methane production (Papadopulos and Associates and Colorado Geological Survey, 2007).

Source: Jenkins (1968), Kansas Department of Agriculture (2010), Nebraska DNR (2004; 2007), Papadoulos and Associates and Colorado Geological Survey (2007).

Using the Glover-Balmer equation to solve the above defined problem leads to an optimal solution for which *the ratio of the marginal benefit from pumping to the marginal externality caused by pumping should be equal across all well locations*. This common ratio may then be interpreted as the effective (present value) optimal entitlement price. In an agricultural groundwater use setting, Kuwayama and Brozović (2013) have further shown that if the marginal damage of the externality is the same for all farms, this outcome can be induced with marketable entitlements that are traded on a one-to-one basis, where the marginal abatement costs of all firms will equal marginal damage multiplied by this ratio. Conversely, if the marginal benefit function is the same at each pumping location, then wells closer to the stream will always be more constrained than wells further from the stream.

Such models can also be used for designing groundwater regulations. Annex 3.B provides a complete case study of an applied economic model used to compare the cost-effectiveness of alternative policy instruments, such as land retirement, quotas, or tradable permits to address simulation. The presented economic solutions provide a good basis on which to gauge the advantage and drawbacks of policy instruments, which will be the object of the following two sections (demand-side and supply-side approaches).

Demand-side policy instruments to manage groundwater use

An important point from the literature is that no single policy instrument is "superior" to others; the choice depends on a good understanding of localised hydrology, institutions, and the specific kinds of externalities that the policy maker wishes to address (Kuwayama and Brozović, 2013; see also Annex 3.B). Moreover, even in one location, the choice of instrument will also depend on the desired level of reduction in water use: policy rankings are not necessarily invariant to changes in total water use or desired hydrologic conditions (Palazzo and Brozović, 2014).

The following sub-sections review the core management instruments under the three broad categories defined in Table 3.1: regulatory, economic, and collective management. While presented individually for clarity purposes, it should be noted that several of these are applied in combination.

Other important related conditions are then addressed, such as monitoring and enforcement and policy coherence with other instruments.

Regulatory instruments: Pumping entitlements, quotas and zoning

Groundwater management is generally supported by regulatory instruments. One core approach is to allocate allowable groundwater pumping to each well. These allocations or entitlements may be based on the historical irrigated area and expected crop water demands, or on historical usage. Allocations for each unit of irrigated area may be set equally, or different allocations may be set according to hydrologic zone or irrigation technology used.

From an economic point of view — following results from the above presented model — uniform quantitative restrictions (or quotas) are not a cost-effective method to reach any hydrologic goals unless producers have identical benefit functions and impacts on the surface water-groundwater system. This is because, in general, providing each user the same allocation will not equate the marginal benefits of water use, as specified in the model developed above. However, quotas are generally viewed as an equitable regulation as they are imposed equally on entitlement holders.[11] As such, if there is expected to be little spatial variation in the impacts of producers on the hydrologic system, and if benefit functions are also expected to be similar, then a uniform quota may be an effective groundwater management tool. As the hydrologic complexity and producer heterogeneity increases, quotas need to be targeted progressively with increasing care and information requirements for effective regulation will increase quickly.

Zoning is another type of regulatory instrument for groundwater management that restricts certain kinds of activities within defined areas. For example, pumping may be limited within a certain distance of a stream that contains critical habitat, or within a certain distance of a town's drinking water supply. Well spacing requirements that limit well density are also a type of zoning regulation. Zoning is commonly decided by hydrologists, geologists, or environmental engineers working for a government regulatory agency, and is thus often sensitive to local hydrologic conditions. For example, well spacing restrictions are often set explicitly on the basis of local hydrologic properties to avoid significant well interference. Thus, above aquifers with higher transmissivities and lower storativities, there are correspondingly larger well-spacing requirements (e.g. see Brozović et al., 2010).

Economic instruments: Redressing farmers' incentives

Economic instruments differ from regulatory instruments in that they do not place absolute limits on producer behaviour, but provide a price signal whose intent is to encourage producers to change their behaviour in a way that achieves the desired hydrologic outcome. Such incentive-based schemes may produce revenue for the regulator (taxes), may be costly to the regulator (subsidies), or may involve payments only between producers (trading). The broad categories of instruments considered here are taxes (whether on pumped water or on a proxy), transferrable entitlement systems, and buy-outs of land or water entitlements.

An important difference between the various kinds of economic instruments is the extent to which widespread and enforceable groundwater monitoring is necessary. Instruments that provide incentives related to the physical quantity of groundwater used, such as pumping taxes and some transferable permit systems, require the ability to monitor water use accurately and to confirm that changes in water use are occurring as a result of the incentive. Instruments that provide incentives related to the quantity of land in irrigation may have lower hurdles for implementation and enforcement as there is typically already a monitoring system in place related to assessing value and property taxation.

In areas where there are no binding and enforced groundwater use regulations, incentive-based instruments will be difficult to implement and their outcome will be hard to quantify. Thus, when considering alternate policy instruments for groundwater management from a practical perspective, it is important to understand that incentive-based instruments are often used in conjunction with underlying regulatory approaches that provide monitoring and enforcement, rather than independently.

Pricing: The efficiency-acceptability trade-off of taxes and subsidies

From an economic modelling perspective, it is straightforward to show that a producer facing a unit tax on groundwater that is used as a production input will choose to use an amount at which the marginal benefit of water use (i.e. the value of the marginal product of water) is equal to the tax. It follows that a tax that is chosen to meet the optimality conditions developed in the model will be a cost-effective instrument to attain any hydrologic target. If the regulator is primarily concerned with the total amount of pumping from the aquifer, the tax will be uniform. If there are spatial concerns related to desired hydrologic conditions, the tax will also vary spatially and be normalised according to hydrologic impact. Per-unit taxes will be larger for wells that have a higher impact on the surface water-groundwater system.

From a regulator's perspective, taxes may be desirable as they are both potentially an optimal solution to the management problem and they may be revenue-generating, but there are important caveats.[12] First, it may be very unpopular with producers to introduce new taxes on an input used in agricultural production and not all management institutions have the legal authority to do so. The second shortcoming of taxes is that the demand for irrigation water can be quite inelastic (e.g. Koundouri, 2004; Schoengold et al., 2006).[13] Hendricks and Peterson (2012), focusing on the state of Kansas where groundwater is pumped for irrigation from the High Plains Aquifer, find an estimate of -0.1, with responses occurring on the intensive margin, pointing out the possible inflexibility of farmers to respond to new conditions. Zhu et al. (2012) also consider that -0.10 is a reasonable average elasticity for groundwater use in agriculture at the global level. Several studies also point to the lack of evidence of the effect of taxing on groundwater use (e.g. EEA, 2013).[14] These results mean that, generally speaking, high per-unit pumping taxes may be needed to change pumping behaviour meaningfully in order to achieve significant changes in hydrologic impact. High taxes may not, however, be politically feasible.

Even if it is not possible to tax water use, it may be possible to tax some other input related to water in the production process. One possibility is to tax irrigated land on a per-area basis. Such taxes are sometimes referred to as occupation taxes. Here again, a high level of taxes may be needed (e.g. Schoengold et al., 2006; Hendricks and Peterson, 2012), but politically this may be very unpopular.

Another possibility is to use indirect means by taxing energy to reduce groundwater pumping, but the results are not guaranteed either. A global simulation showed that doubling energy prices would result in only limited reduction in groundwater use (-7.5% in groundwater depletion, see Zhu et al., 2012). In the United States, Hendricks and Peterson (2012) found that increasing the price of energy (e.g. via taxing) would not, on average, be an effective tool to manage groundwater use over the High Plains Aquifer. Still, Pfeiffer and Lin (2014), focusing on a specific area of this major aquifer and looking more specifically at electricity prices, found elasticities to be higher in absolute value (-0.26) than the previously cited references, suggesting that an increase in prices in that region would actually result in changes in behaviour leading to lower agricultural groundwater use (more on energy pricing is described in the section on intersection of policies).[15]

Subsidies to support reductions in groundwater use may have a similar aggregate environmental and welfare impact as taxes, but involve transfers of funds from the regulator to producers.

Unsurprisingly, subsidies are much more politically acceptable to producers than taxes. As currently structured, most subsidy programs in groundwater management do not directly subsidise reductions in pumping, but offer cost-share incentives to producers to undertake new management practices that will reduce consumptive water use (e.g. UNL Extension, 2014). For example, cost sharing for soil moisture-sensing technology or for irrigation technology updates are examples of subsidies. In such cases, targeting the right users and objectives can be critical, even challenging, in situations of information asymmetries. Even if well targeted, subsidy programs can be effective but they are also costly to the regulator. Moreover, if fine-scale spatial targeting is needed to achieve desired environmental goals, subsidies may be difficult to implement as they typically involve a voluntary sign-up to the incentive.

Subsidies for irrigation efficiency, as found in Australia, can act as a double-edged sword in groundwater management. Increasing efficiency can be beneficial for groundwater resources and agriculture (Pacific Institute and NRDC, 2014), but it reduces recharge and may result in additional water uses without external constraints. First, subsidies for irrigation efficiency have been found to increase water use as higher crop yields lead to higher evapotranspiration with no return flow or recharge in aquifers (OECD, 2012a). In addition, behavioural responses can also make such programs ineffective. A program supporting irrigation efficiency in a groundwater irrigated area over the High Plains Aquifer in western Kansas pushed farmers to extend planting and switch crops towards more water intensive options, leading to higher water use overall (Pfeiffer and Lin, 2014). Linking efficiency with overall quotas on withdrawals can prevent some of these effects. Restrictions on withdrawals, by themselves, can also encourage water savings and increase irrigation efficiency.

It should be noted that tax exemptions (or incomplete pricing) for irrigation water can be considered an implicit groundwater use subsidy, and not accounting for scarcity and externality can be translated as decreasing the marginal opportunity cost of water, encouraging inefficient use of groundwater. In the Netherlands, for instance, farmers are exempted from a groundwater tax (up to a certain threshold), which has encouraged them to use multiple smaller pumps to avoid the tax (OECD, 2008).

Groundwater markets: Cost effective but involves transaction costs

In areas where regulatory restrictions on groundwater withdrawals are already in place, it may be possible to introduce instruments that allow producers to transfer the entitlements they currently have amongst themselves. Such market systems are a cost-effective method to achieve any given water use reduction or hydrologic target as they allow equalisation of the values of marginal products (Kuwayama and Brozović, 2013; Palazzo and Brozović, 2014). There are, however, significant preconditions to a successful groundwater market: a strong property rights system, a robust price determination mechanism with information, and infrastructures are required (Skurray et al., 2013).

In aggregate and in the absence of transaction costs, transferable entitlement systems can achieve an outcome identical to tax instruments because in principle and in the absence of spatial complexity, both transferable entitlement systems and tax instruments equalise marginal abatement costs (with or without consideration of marginal externalities) across all water users (Montgomery 1972; Kuwayama and Brozović, 2013). Although they may not generate revenue for the regulator,[16] they do result in transfers of funds between buyers and sellers of water entitlements. If the regulatory goal is to reduce aggregate water use, the entitlement system will have a single market-clearing price that is equal to a tax that would achieve the same aggregate pumping. In this case, the marginal benefits of water are equalised across all traders, and a frictionless transferable entitlement scheme will by definition achieve the optimal allocation (e.g. Montgomery 1972; Sunding et al., 2002; Jaeger 2004). The more binding the constraint on total water used, the larger the equilibrium entitlement price. If the regulatory goal is to address a spatial externality, then the price of entitlements in the water market will be adjusted so that the ratio of the marginal benefit to the marginal externality is equal across all trading locations. In the latter case, there may not be one market-clearing price.

Transferable entitlement schemes are voluntary and will benefit both buyers and sellers, but the level of transaction costs matters. In the High Plains region of the United States, there are examples of transferable entitlement systems that have been either relatively successful or quite unsuccessful; one difference has been the role of transaction costs (Brozović and Young, 2014). If monitoring and enforcement of water entitlements are already in place, then the transaction costs of introducing a transferable entitlement scheme may be low. However, if monitoring and enforcement must first be introduced, then the entitlement system may be expensive to administer and may also face political opposition from stakeholders. The number of traders will affect the transaction costs, but all relevant users should be included to avoid leakages. Furthermore, legal, institutional, and environmental barriers may need to be overcome (Garrido et al., 2012).[17]

Even if groundwater metering is not present, it may be possible to establish a transferable entitlement system for groundwater. For example, if the total irrigated area within a water district is constrained to be less than the total area potentially available for cropland within a district, then the right to irrigate units of land can be reallocated using an entitlement system. However, for a land-based permit system, the aggregate pumping resulting after transfers will be subject to uncertainty. This may be acceptable if the expected variation in marginal externality is much larger than the expected variation in water application rates (Young, 2014). However, there is still a need to monitor and enforce limits on irrigated area for such systems to succeed.

Indirect control: Irrigation entitlements retirements is a conservation tool whose cost depends on targeting

A land retirement program operates through existing land markets. Farmland with the right to irrigate is purchased and then the irrigation entitlements are retired. Formerly irrigated land moves to dryland (i.e. rain fed only) agriculture. Thus, aggregate water use within the groundwater management area is reduced by an amount equal to the total pumping entitlements associated with the purchased land. Or the irrigation entitlement may be purchased and retired by itself, separately from the land. From an economic point of view, land or water retirement programs will be a relatively expensive solution as they generally operate on the extensive and not the intensive margin (area of irrigation rather than groundwater use on the same land).[18] Moreover, a limited range of entitlements may be available for acquisition at any point in time.

However, the transaction costs of entitlement retirement may be low as only one landowner at a time is involved. Entitlement retirement programs may be targeted based on the cheapest land (reducing irrigated acreage the fastest way possible), the cheapest water (reducing aggregate pumping the fastest way possible), or by impact on the hydrologic system (shutting down wells with the highest marginal externalities as quickly as possible). Each of these targeting options will have different individual and aggregate impacts at a different cost. If the primary goal of regulation is to address a spatial externality within a surface water-groundwater system, such as stream depletion, then targeting land retirement based on that externality (i.e. paying a premium to purchase entitlements that have higher associated marginal externalities) will be cost-effective. Note however, that if there is a strong positive correlation between the marginal benefits of groundwater use and the marginal externality of groundwater use, the optimal solution may be to retire a relatively large area of less hydrologically-damaging but low-cost land.

Collective management approaches: Locally adapted but reliant on stakeholder participation

The third category of instruments addressing demand-side challenges is the broad category of collective management schemes. This includes voluntary programs undertaken by users to manage or reduce groundwater use, or collective mechanisms induced by regulatory framework requirements. It is possible for local groundwater users' groups to introduce binding management policies collectively, voluntarily, often in conjunction with other higher level instruments, or sometimes set up independently.

There are several potential concerns that can prompt user groups to develop voluntary instruments (Lopez-Gunn and Martinez Cortina, 2006). First, in some regions, stakeholders have long-standing concerns over the ability of future generations to continue profitable irrigated agriculture and local governance structures are adaptable enough to allow for the implementation of self-regulation and self-enforcement to reduce aggregate pumping. Second, self-regulation can ease the implementation of regulations to help to prevent externalities. Third, there are local groundwater management areas where there is a real possibility of stringent regulations being imposed from a regional or federal administrative level, typically in response to impacts on endangered species habitat or groundwater-dependent ecosystems. In some such cases, local management bodies have tried to introduce regulations voluntarily in a "pre-compliance" setting, where one goal is to try to achieve desirable environmental outcomes while maintaining local control. Where groundwater management policies are introduced voluntarily, they generally are of the kind already described, and issues of monitoring and enforcement are equally present.

One of the advantages of collective management approaches is their scope of action, close to the aquifer, and therefore able to incorporate specific challenges and users specificities. An interpretation of this property under the modelling scheme is that collective management schemes allow constraints to be adapted to users' own challenges and capacities to some extent which may avoid costly mistakes due to the heterogeneity of individual farmers' situations. Such internalisation of the constraints can avoid costly and challenging targeting exercises, and may encourage individual farmers to take action. This does not mean that a framing regulation is not necessary or that monitoring will operate in a voluntary manner, but some of the transaction costs encountered with other instruments will be shifted to the collective of operators, and may remain lower overall.

Other related conditions for effective groundwater management: Enforcement and policy alignment

In addition to the choice of policy instrument, there are several other concerns that need to be addressed in the design and implementation of groundwater management policies. Some of these, such as the need to monitor and enforce restrictions, are common to all water regulations and to environmental regulations in general. Others, such as the need to understand local and regional hydrology and how these control groundwater flow, recharge, and surface water-groundwater interaction, are specific to groundwater management. It is always necessary to consider whether there are other policies, such as agriculture or energy policies, whose primary intent may not be to manage water use but which nevertheless impact individual and group decision-making about water use in unanticipated ways.

Monitoring and enforcement: A key necessary condition for functional policy frameworks

A core consideration is the capacity of monitoring and information systems that precede any policy action (e.g. Mechlem, 2012; Morris et al., 2003; Struzik, 2013). Groundwater is a largely invisible resource, and can therefore remain out of any scope of action even in the case of justifiable and required public intervention. Monitoring can take different forms, via direct or indirect instruments, and uses. It may encompass monitoring of the reserves, flows, quality and the interactions with surface water bodies and can be undertaken by private individual actors and/or by local, regional or national entities.

It is important to note that all groundwater management policies are only effective if they are accompanied by credible monitoring and enforcement of violations. In many regions of OECD countries, there is neither effective monitoring of groundwater use in agriculture nor a way to credibly enforce restrictions of groundwater use.[19] As a result, there is no straightforward way to implement policy that will change individual or group behaviour.

Monitoring of groundwater use is only meaningful to resource management to the extent that there is enforcement when violations occur. Where reporting of meter data is voluntary and without sanction, there is little incentive to provide timely or accurate readings. Conversely, in some groundwater management areas, paid employees undertake water meter reading, with fines for broken meters and severe penalties for violators.

It should be noted that even if well metering is not present and is politically unacceptable to implement, imperfect monitoring may be sufficient to establish regulations. For example, it may be possible to regulate a proxy for the volume of water used in agriculture, such as the area of irrigated land. If per-unit area irrigation applications do not vary by a large amount (i.e. if similar irrigation technologies and cropping practices are used throughout the area to be regulated), then the uncertainty introduced by lack of metering may be quite small. Electricity records may be used as a proxy for groundwater pumping, or historically-irrigated areas may be certified as the only ones that can be irrigated using groundwater. Even with imperfect monitoring, there is still a need to enforce limits on the irrigated areas (or other proxy used) for such systems to succeed.

Interaction with other policies: Aligning energy and agriculture policies

It is important to consider how groundwater management policies may interact with other policies that influence decision-making about crops. In some cases, other policies may unintentionally be the major drivers of decisions about water use. Four examples are provided here.

First, energy pricing policies may have consequences on groundwater pumping decisions (Scott, 2013; Mieno, 2014). India and Mexico are known to have applied high energy subsidies on electricity for agricultural users. By reducing marginal costs of groundwater use, these policies have increased the incentive to pump groundwater. Multiple report have found that these subsidies have increased water use, leading to high energy use and financial costs, with very limited benefits for farmers (OECD, 2008; OECD, 2010; OECD, 2012b). As shown in Table 3.2, recent studies have shown that reducing or removing these subsidies would result in significant reduced groundwater uses there.

Table 3.2. Linking energy policies and groundwater use

Region	Price demand elasticity of water or energy	Implication	Source
India	-0.13 (water)	10% reduction subsidy → 4.4% reduction in water extraction	Badiani and Jessoe (2011) Badiani et al. (2012)
Mexico	n/a	Eliminating electricity subsidy would lead to 15% less pumping in the short run, 19% long term	OECD (2013b)

Source: Author's own compilation, based on Badiani and Jessoe (2011); Badiani et al. (2012), and OECD (2013b).

Second, related concerns over peak load management by rural energy providers have led to many producers opting in to energy supplies which may be cut off during times of high demand, but are also much cheaper per unit of energy than non-interruptible supplies. However, because interruptible supplies induce producers to irrigate at times that are not optimal from the perspective of crop evapotranspiration, they may provide an incentive to over-apply water when applications occur. In such cases, there may be a fundamental tension between the goals of energy management and those of groundwater management (Mieno, 2014).

Third, current biofuel policies in the United States may be encouraging increased groundwater use through a number of mechanisms, including increased commodity prices, the water demands of ethanol processing facilities, and both intensive and extensive margin effects of biofuel feedstock on irrigated land (Schaible and Aillery, 2012).

Lastly, agricultural insurance programs can affect the use of groundwater, In the United States, many crop insurance contracts require producers to irrigate a crop until the end of the growing season to be eligible for payments, even if the crop has already failed. This clearly is problematic both for economic and water conservation reasons. More broadly, drought insurance programs can act as adaptation instruments, but if improperly priced, may also result in increased groundwater use. By decoupling the water needs from income flows, drought insurance may result in lowering the incentive to adapt, and therefore prevent saving groundwater in the long run. As such, subsidies that support crop insurance comprising irrigated crops may result in additional groundwater use in specific areas. Still, the relationship between crop insurance and irrigation water use remains unclear (2014 OECD questionnaire).

Agricultural income support programs may also encourage groundwater use, especially if they support the production of water intensive commodities (like corn) in groundwater irrigated areas. These subsidies will result in lower opportunity cost of water, including groundwater, leading to sub-optimal groundwater use.

Supply-side approaches: Relieving the constraints for users, at a cost

Supply-side approaches consist in increasing water available to farmers either in the short run via alternative (surface water) supplies or in the long run by storing groundwater. Accessing this additional water is intended to lessen pressure on aquifers. By increasing the supply of surface water or groundwater, they are designed to relieve the constraint for managers and users, and therefore relax or delay the above presented optimisation problem, rather than attempting to solve it (Lopez-Gunn et al., 2012). Still, they can contribute as complements to other management schemes, and release binding constraints when facing serious scarcity problems and associated externalities, such as salinity or land subsidence (chapter 2).

As such, supply-side approaches do not support the management of groundwater use *per se* and should therefore not be prioritized. They do not affect agricultural systems in the same way as demand-side approach and may not be equally efficient (Lopez-Gunn et al., 2012a). Demand-side approaches will increase resilience of the production system to shocks while supply side approaches, on their own, may result in disincentives to take action to limit pumping (OECD, 2010).

Whether via supplementing or storage, the supply-side approaches use surface water as a backstop for groundwater, and can be considered under the broader spectrum of conjunctive surface-groundwater management (Box 2.3). The main principles of conjunctive management are simple: groundwater serves as a support in cases where surface water is insufficient, and conversely surface water is used to replenish groundwater resources. In cases where the water table increases too much, surface water use is replaced by groundwater pumping (Ribeiro and da Cunha, 2010).

The economic viability of such approaches critically depends on fixed costs. Still, not all of these instruments require the same degree of public investment. Rainwater harvesting can be set up by farmers on their own, and infiltration ponds can be introduced by groups of farmers or local districts. In contrast, water reservoir expansion and desalination require relatively heavy infrastructure and variable costs. The scale of aquifer storage and recovery and groundwater banking is significant in its cost, but past experience has shown that local collective management schemes have been able to self-finance and operate such programs.

Synthesizing lessons from the economics literature: A call for adaptive management policies

As shown in the synthesis presented in Table 3.3, each of the policy instruments reviewed has advantages and drawbacks; there is no simple economic ranking across instruments. Even within a single region, the welfare rankings of alternative policies may vary depending on the desired environmental and hydrological goals and context.

Table 3.3. From economics to policy: Comparing instruments to manage groundwater

Main approach		Instrument	Advantages	Drawbacks	Factors conditioning success
Demand-side	Regulatory instruments	Entitlements	Core measure to control groundwater use	Depends on allocation mechanism	Flexible and adaptable allocative rules
		Uniform quotas	Equitable	Not cost-effective	Limited spatial complexity
		Zoning	Cope with well interference	Sensitive to local hydrological conditions	Requires scientific expertise
	Economic instruments	Taxes (groundwater)	Optimal solution and revenue generating	Ineffective at low levels, unpopular	Expertise required to set and adjust levels
		Taxes on land	Replacement solution, revenue generating	Second-best, ineffective at low levels, unpopular	Idem
		Taxes on energy	Replacement solution, revenue generating	Generally ineffective, may depend on energy market	Idem
		Subsidies (cost share)	Acceptable, effective solutions	Costly and difficult to implement, voluntary	Designing incentives to participate
		Support for Irrigation efficiency	Long term reduction in consumption	Reduces recharge, risk of rebound effect, costly	Works better with overall quotas
		Groundwater markets	Cost-effective optimal solution	Transaction costs	Significant preconditions
		Land-based transfers	Second-best solution	Does not guarantee results	
		Irrigation entitlement retirements	Second-best solution	Can be costly and ineffective	Lower transaction costs via targeting
	Collective management approaches	Voluntary programs	Internalise local constraints Lower transaction costs	Adoption and implementation dependent (risk of free riders)	Overarching regulatory framework

Table 3.3. From economics to policy: Comparing instruments to manage groundwater (*cont.*)

Main approach		Instrument	Advantages	Drawbacks	Factors condition success
Supply-side	Alternative supplies	Rainwater harvesting	Low investment	Low results, weather dependent	
		Water reservoir expansion	Relieve constraints	High costs and investments needed, possible ecosystem damages, weather dependent	Long-term investment
		Desalination	Relieve constraints	High costs, energy dependent, possible ecosystem damages	Long-term investment
	Groundwater Storage	Infiltration ponds	Low cost recharge	Recharge rates may vary	Expertise
		Aquifer storage and recovery	Relieve constraints and encourage recycling	High costs and uncertainty of results	Expertise and financing
		Groundwater banking			

This suggests that policy makers should focus on supporting adaptive management of groundwater resources. This can be done in two ways. First, they should support enabling factors for multiple regulatory instruments: increasing information collection and analysis, improving monitoring and enforcement, lowering transaction costs, and supporting financing schemes. Second, they will need to support locally customised options, for instance building on existing regulatory frameworks with engagement from stakeholders, and combining instruments that work better together.

Chapters 4 and 5 will complement this analysis with reviews and assessments of policies in OECD countries to move towards recommendations.

Notes

1. In particular, several illustrative examples in this chapter are drawn from areas located at different locations of the High Plains Aquifer in the United States, notably because of the wide diversity of management schemes applied in this region, the diversity of climate it covers, and because of the explicit reliance, for some of the local groundwater management districts, on economic analysis as a basis for policy design. This does not mean, however, that groundwater systems there are equivalent to others.
2. See glossary for definitions.
3. It should be noted that in cases where wells that are operated by different producers are closely spaced, or where local hydrology means that small changes in saturated thickness lead to large changes in well yield and available pumping rates, well interference may be a concern (e.g. Saak and Peterson, 2007).
4. Some experts focus on whether solutions are technological or institutional (Giordano, 2009). Others compare direct versus indirect approaches (Kemper, 2007), consider the legal status of groundwater (Llamas et al., 2007), the main governance instruments (Foster et al., 2009; Custodio, 2010), or the type of institutions: state-based or public, market-based or private, and collective or user-based (Meinzen-Dick, 2007). Shah et al. (2008) characterise four direct instruments used in

groundwater management: direct administrative regulations, economic instruments, tradable water entitlements, and participatory and participatory aquifer management approaches

5. In areas where aquifers have high hydrologic connectivity and relatively low-yielding wells, intraseasonal groundwater elevation changes may be of concern. However, the analytical focus of this chapter is on current pumping levels, with surface impacts as the relevant constraints. This focus matches most current regulations, which generally do not explicitly model either well interference or longer-term aquifer dynamics. Implicitly, this means that the medium-term dynamics that are relevant for surface water-groundwater interaction are accounted for in the model presented.

6. This is consistent with well spacing observed in many OECD countries (e.g. Brozović et al., 2010). However, in areas where only shallow irrigation wells are in place and groundwater demands are rising rapidly, such as in areas that have traditionally not relied heavily on groundwater to meet their crop evapotranspiration requirements, well interference may create issues during prolonged droughts.

7. This vector may impact the hydrologic constraints (e.g. If climate becomes drier, then aquifer recharge in the long run will be reduced). However, this kind of dynamic does not enter optimal decision making in the short run.

8. Surface water availability is not explicit but embedded into the objective via the crop production function. The user will not pump if its crop does not demand it. The objective can also respond to environmental effects, such as ecosystem damages, so long as the necessary water restriction can be defined.

9. An alternate approach that may fit some regulatory settings better is the "dual" one, where the objective is to minimise overall depletion subject to net benefits equal to or exceeding the current ones.

10. One manifestation of this variability is the observed volatility of spot market prices for water in thick surface water markets, such as those in the Murray-Darling Basin in Australia.

11. Groundwater use is generally defined based on the use of licenses, permits or rights. To avoid focusing on one or the other, the term "entitlement", under a broad definition (see glossary for detail), will be used in this report.

12. Some taxes can be compensated and therefore do not generate public revenue.

13. General estimates of price elasticities for irrigation demand (including surface water) may vary widely. Scheierling et al. (2006) review 24 estimates of the price demand elasticity of irrigated water and find a range between -0.001 to -1.97. Irrigation demand is generally inelastic below a price threshold and elastic above (Koundouri, 2004). The threshold depends on climatic conditions; it is higher in dry season. Water scarcity level also matters in the elasticities; water-scarce regions will be especially inelastic. And so will revenues: high value crops are associated with inelastic irrigation demand (FAO, 2011).

14. Another case for which groundwater taxing will not be effective is that of irrigation systems with conjunctive use of surface and groundwater, if the adjustment is not also borne by surface water (Schuerhoff et al., 2013).

15. An increase of energy prices of USD 1 per million British thermal unit was found to reduce groundwater pumping by 3.6% (Pfeiffer and Lin, 2014).

16. This is true except when the regulator auctions the permits. Regulators can also pay for rights, e.g. via environmental buy-outs.

17. Legal barriers include market barriers (instituted from monopolies or public services) and those related to the definition of water entitlements. Institutional barriers to trade are regionally based on intersectoral concerns due to certain actors opposing trades. Environmental barriers may be set up by agencies in charge of ecosystems and water quality (Garrido et al., 2012).

18. In the context of groundwater-fed irrigation, the intensive margin decision refers to the per-area irrigation intensity chosen during the irrigation season. The extensive margin decision refers to the choice of total area to be irrigated, which for an annual crop is a planting decision that occurs before the irrigation season

19. Illegal wells are frequent in some of the large agricultural groundwater-using regions, thus preventing other groundwater measures to be effective in reducing pressure on an aquifer (see Chapter 4 for details).

References

Athanassoglou, S et al. (2012), "Optimal Mechanisms for Heterogeneous Multi-Cell Aquifers", *Environmental and Resource Economics*, Vol. 52:265–291.

Badiani, R. and K.K. Jessoe (2013), "The impact of electricity subsidies on groundwater extraction and agricultural production", Department of Agriculture and Resource Economics Working Paper, University of California Davis.

Badiani, R., K.K. Jessoe, and S. Plant (2012). "Development and the Environment: The Implications of Agricultural Electricity Subsidies in India", *The Journal of Environment & Development*, Vol. 21(2), pp. 244–262.

Brozović, N., D.L. Sunding, and D. Zilberman (2010), "On the spatial nature of the groundwater pumping externality", *Resource and Energy Economics*, Vol. 32(2), pp. 154-164.

Brozović, N., D.L. Sunding and D. Zilberman (2006), "Optimal management of groundwater over space and time", in R. U. Goetz and D. Berga (eds.), *Frontiers in water resource economics*, Springer, New York.

Brozović, N. and R. Young (2014), "Design and Implementation of Markets for Groundwater Pumping Rights", in K. Easter and Q. Huang (eds.), *Water Markets for the 21st Century: What Have We Learned?*, Springer, New York.

Custodio, E. (2010), "Intensive groundwater development: a water cycle transformation, a social revolution, a management challenge", in L. Martinez Cortina, A. Garrido, and E. Lopez-Gunn (Eds.), *Water Workshop on Re – thinking Water and Food Security Paradigms*, CRC Press, London.

Dixon, L. (1989), "Models of groundwater extraction with an examination of agricultural water use in Kern County", California, Ph.D. thesis, University of California, Berkeley.

Domenico, P.A. (1972), *Concepts and Models in Groundwater Hydrology*, McGraw-Hill, New York.

Donohew, Z. (2013), "Markets and private property rights to groundwater", *Conservation Leadership Council Policy Paper*, July 2013, 17 pp.

English, M. (1990), "Deficit Irrigation I: Analytical Framework", *Journal of Irrigation and Drainage* Engineering, Vol. 116 (3), pp. 399–412.

EEA (European Environmental Agency) (2013), "Assessment of cost recovery through water pricing", EEA, Copenhagen.

Fanning, J. (2012), Direct testimony of Jasper Fanning, Manager of the Upper Republican Natural Resources District, in the Supreme Court of the United States, State of Kansas (Plaintiff) v. State of Nebraska and State of Colorado (Defendants), No. 126.

FAO (United Nations Food and Agricultural Organization) (2011), *The State of the World's Land and Water Resources for Food and Agriculture: Managing systems at risk*, Food and Agricultural Organization Publications, Rome.

Foster et al. (2009), "Sustainable Groundwater Management Contributions to Policy Promotion", GW-MATE, Strategic Overview Series 1, The World Bank, Washington, DC.

Freeze, A. and J. Cherry (1979), *Groundwater*, Prentice-Hall, Englewood Cliffs, NJ.

Gardũno, H. and S. Foster (2010), "Sustainable groundwater irrigation: approaches to reconciling demand with resources", GW-MATE, Strategic Overview Series 4, The World Bank, Washington, DC.

Garrido, A., D. Rey, and J. Calatrava (2012), "Water trading in Spain", in de Stefano and R.M. Llamas (eds.), *Water, Agriculture and the Environment in Spain: Can we square the circle?,* CRC Press, London.

Giordano, M. (2009) "Global Groundwater? Issues and Solutions", *Annual Review of Environment and Resources*, Vol. 34(1), pp. 153–178.

Gisser, M., and Sanchez, D. (1980), "Competition versus optimal control in groundwater pumping", *Water Resources Research*, Vol. 16 (4), pp. 638-642.

Glover, R., and Balmer, G. (1954), "River depletion resulting from pumping a well near a river", American *Geophysical Union Transactions*, Vol. 35 (3), pp. 468–470.

GWP (Global Water Partnership) (2012), "Groundwater Resources and Irrigated Agriculture", GWP, Stockholm Sweden.

Han, J. (2011), "Spatial dynamic groundwater regulation to ensure adequate instream flow for salmon", unpublished MS thesis, University of Illinois at Urbana-Champaign, Department of Agricultural and Consumer Economics, Urbana.

Hendricks, N.P., and J.M. Peterson (2012), "Fixed Effects Estimation of the Intensive and Extensive Margins of Irrigation Water Demand", *Journal of Agricultural and Resource Economics*, Vol. 37 (1), pp. 1-19.

Huang, Q., et al. (2012) "The Effects of Well Management and the Nature of the Aquifer on Groundwater Resources", *American Journal of Agricultural Economics*, Vol. 95(1), pp. 94–116.

Hunt, B. (2012), "Review of stream depletion solutions, behavior, and calculations", Journal of Hydrologic Engineering.

Hunt, B. (1999), "Unsteady stream depletion from ground water pumping", *Ground Water*, Vol. 37(1), pp. 98-102.

Jaeger, W. K. (2004), "Conflicts over water in the Upper Klamath Basin and the potential role for market-based allocations", *Journal of Agricultural and Resource Economics*, 29, 167–184.

Jenkins, C.T. (1968), "Techniques for computing rate and volume of stream depletion by wells", *Ground Water*, Vol. 6, 37–46.

Kansas Department of Agriculture (2010), *Rules and Regulations: Kansas Water Appropriation Act*, Topeka, Kansas.

Kemper, K.E. (2007), "Instruments and institutions for groundwater management" in Giordano and Villholth (eds.), *The Agricultural Groundwater Revolution: Opportunities and Threats to Development*, CABI, Oxford.

Konikow, L.F. (2013), Groundwater depletion in the United States (1900–2008): U.S. Geological Survey Scientific Investigations Report 2013–5079, 63 p., http://pubs.usgs.gov/sir/2013/5079.

Koundouri, P. (2004), "Current Issues in the Economics of Groundwater Resource Management", Journal of Economic Surveys, Vol. 18(5), 703–740.

Kuwayama, Y., and Brozović, N. (2013), "The regulation of a spatially heterogeneous externality: tradable groundwater permits to protect instream flows", Journal of Environmental Economics and Management, Vol. 66(2), 364–382.

Llamas, R. (2004), "Water and Ethics. Use of groundwater", UNESCO Series on Water and Ethics, Essay, UNESCO, Paris.

Llamas, R.M. and A. Garrido (2007), "Lessons from intensive groundwater use in Spain: Economic and social benefits and conflicts", In Giordano, M. and K.G. Villholth (eds.), *The Agricultural Groundwater Revolution: Opportunities and Threats to Development*, CABI Oxford.

Llamas, R.M., and P. Martínez-Santos (2005), "Intensive groundwater use: silent revolution and potential source of social conflicts", *Journal of Water Resources Planning and Management,* Vol. 337–342.

Lopez-Gunn, E., and L. Martinez Cortina (2006), "Is self-regulation a myth? Case study on Spanish groundwater user associations and the role of higher-level authorities", *Hydrogeology Journal*, Vol. 14(3), 361–379. doi:10.1007/s10040-005-0014-z

Lopez-Gunn et al. (2012), "Taming the groundwater chaos", in de Stefano, L. and M.R. Llamas (eds.), *Water, Agriculture and the Environment in Spain: Can we square the circle?*, CRC Press, London.

Madani, K., and A. Dinar (2013), "Exogenous Regulatory Institutions for Sustainable Common Pool Resource Management: Application to Groundwater", *Water Resources and Economics*, Vol. 2-3, pp. 57-76.

Martin, D., et al. (2007), "Water Optimizer: a decision support tool for producers with limited water", University of Nebraska-Lincoln Departments of Biological Systems Engineering and Agricultural Economics.

McKusick, V. (2002), "State of Kansas v. State of Nebraska and State of Colorado: Joint Motion of the States for Entry of Proposed Consent Judgement and Approval and Adoption of Final Settlement Stipulation", Supreme Court of the United States, 2002.

Mechlem, K. (2012), "Legal and Institutional Frameworks", Thematic Paper 6, Groundwater Governance: A Global Framework for Action, GEF, FAO, UNESCO-IHP, IAH, and the World Bank.

Meinzen-Dick, R. (2007), "Beyond panaceas in water institutions", *Proceedings of the National Academy of Sciences*, Vol. 104(39), 15200–15205.

Mieno, T. (2014), "Essays on water resource economics", Unpublished PhD dissertation, University of Illinois at Urbana-Champaign, Department of Agricultural and Consumer Economics, Urbana.

Montginoul, M. and J. Rinaudo (2013), "Quels mécanismes de régulation des prélèvements en eau souterraine? Comparaison du point de vue des agriculteurs, des institutionnels et des citoyens" *Science Eaux et Territoires*, Vol. 11, pp. 64–69. http://www.set-revue.fr/quels-mecanismes-de-regulation-des-prelevements-en-eau-souterraine-comparaison-du-point-de-vue-des-a/texte.

Montgomery, D. (1972), "Markets in licenses and efficient pollution control programs", *Journal of Economic Theory*, Vol. 5 (3), pp. 395–418.

Morris, B. L. et al. (2003), "Groundwater and its Susceptibility to Degradation: A Global Assessment of the Problem and Options for Management", UNEP, Nairobi.

Nebraska DNR (Nebraska Department of Natural Resources) (2007), "Annual Evaluation of Availability of Hydrologically Connected Water Supplies", Lincoln, Nebraska.

Nebraska DNR (Nebraska Department of Natural Resources) (2004), "COHYST Technical Committee. The 40-Year, 28-Percent Stream Depletion Lines for the COHYST Area West of Elm Creek", Lincoln, Nebraska.

Nebraksa DNR and MRNRD (Nebraska Department of Natural Resources and Middle Republican Natural Resources District) (2010), Integrated Management Plan, Lincoln, Nebraska.

Nebraska DNR and URNRD (Nebraska Department of Natural Resources and Upper Republican Natural Resources District) (2010), Integrated Management Plan, Lincoln, Nebraska.

OECD (2015), Water Resources Allocation: Sharing Risks and Opportunities, OECD Studies on Water, OECD Publishing, Paris. http://dx.doi.org/10.1787/9789264229631-en.

OECD (2013a) "Costs and distributional impacts of inaction", in OECD, Water Security for Better Lives, OECD Publishing, Paris.: http://dx.doi.org/10.1787/9789264202405-10-en.

OECD (2013b), *Making Water Reform Happen in Mexico*, OECD Studies on Water, OECD Publishing, Paris. http://dx.doi.org/10.1787/9789264187894-en.

OECD (2012a), "Meeting the water coherence challenge", in OECD, *Meeting the Water Reform Challenge*, OECD Publishing, Paris, http://dx.doi.org/10.1787/9789264170001-7-en.

OECD (2012b), OECD Environmental Outlook to 2050: The Consequences of Inaction, OECD Publishing, Paris, http://dx.doi.org/10.1787/9789264122246-en

OECD (2010), Sustainable Management of Water Resources in Agriculture, OECD Studies on Water, OECD Publishing, Paris, http://dx.doi.org/10.1787/9789264083578-en.

OECD (2008), Environmental Performance of Agriculture in OECD Countries Since 1990, OECD Publishing, Paris, http://dx.doi.org/10.1787/9789264040854-en/.

Pacific Institute and NRDC (Natural Resources Defence Council) (2014), "Agricultural Water Conservation and Efficiency Potential in California", Issue Brief, Oakland, California.

Palazzo, A., and Brozović, N., (2014), "The role of groundwater trading in spatial water management", *Agricultural Water Management*, Vol. 145, pp. 50-60. DOI: 10.1016/j.agwat.2014.03.004.

Papadopulos and Associates and Colorado Geological Survey (2007), "Coalbed Methane Stream Depletion Assessment Study - Raton Basin", *Colorado Geological Survey*, Boulder, Colorado.

Pfeiffer, L., and C.-Y. Lin (2014), "The effects of energy prices on agricultural groundwater extraction from the High Plains Aquifer", *American Journal of Agricultural Economics*, Vol. 96(5), pp. 1349-62.

Roumasset, J.A. and C.A. Wada (2013), "Economics of Groundwater", in *Encyclopedia of Energy, Natural Resource, and Environmental Economics* (1st ed., pp. 10–21), Elsevier.

Saak, A., and J. Peterson 2007), "Groundwater use under incomplete information", *Journal of Environmental Economics and Management*, Vol. 54 (2), pp. 214–228.

Schaible, G., and Aillery, M. (2012), "Water Conservation in Irrigated Agriculture: Trends and Challenges in the Face of Emerging Demands", United States Department of Agriculture Economic Research Service, *Economic Information Bulletin* (EIB-99), Washington DC.

Scheierling, S.M., J.B. Loomis, and R.A. Young (2006), "Irrigation water demand: A meta-analysis of price elasticities", *Water Resources Research*, Vol. 42(1).

Schlager, E. (2007), "Community management of groundwater", in Giordano, M., and K.G. Villholth (eds.), *The Agricultural Groundwater Revolution: Opportunities and Threats to Development*, CABI Oxford, UK.

Schoengold, K., Sunding, D.L., and Moreno, G. (2006), "Price elasticity reconsidered: Panel estimation of an agricultural water demand function", *Water Resources Research*, Vol. 42, W09411, doi:10.1029/2005WR004096.

Schuerhoff, M., H.-P. Weikard and D. Zetland (2013), "The life and death of Dutch groundwater tax", *Water Policy*, Vol. 15(6), 1064. doi:10.2166/wp.2013.112.

Scott, C.A. (2013), "Electricity for groundwater use: constraints and opportunities for adaptive response to climate change", *Environmental Research Letters*, Vol. 8, 035005.

Shah, T. et al. (2008), "Groundwater governance through electricity supply management: Assessing an innovative intervention in Gujarat, western India", *Agricultural Water Management*, Vol. 95(11), pp. 1233–1242.

Skurray, J.H., Roberts, E.J., Pannell, D.J. (2012), "Hydrological challenges to groundwater trading: Lessons from south-west Western Australia", *Journal of Hydrology,* Vol. 412–413, pp. 256-268.

Sophocleous, M. (2002), "Interactions between groundwater and surface water: The state of the science", *Hydrogeology Journal*, Vol. 10, pp. 52–67.

Struzik, E. (2013), "Underground Intelligence: The need to map, monitor, and manage Canada's groundwater resources in an era of drought and climate change", Munk School of Global Affairs, University of Toronto, Toronto, Canada, http://powi.ca/wp-content/uploads/2013/05/POWIUndergroundIntelligence- Struzik-June25.pdf, accessed July 2013

Sunding, D.L., Zilberman, D., Howitt, R., Dinar, A., and MacDougall, N. (2002), "Measuring the costs of Reallocating Water from Agriculture: A Multi-Model Approach", *Natural Resource Modeling*, Vol. 15 (2), pp. 201–225.

Wagner, M.W. and U.P. Kreuter (2004), "Groundwater Supply in Texas: Private Land Considerations in a Rule-of-Capture State", *Society and Natural Resources*, Vol. 17, pp. 349–357.

Young, R.K. (2014), "Tradable permit systems for a spatially heterogeneous externality: A microparameter approach", Master's thesis, University of Illinois at Urbana-Champaign.

Zhu, T., C. Ringler and X. Cai (2007), "Energy price and groundwater extraction for agriculture: Exploring the energy-water-food nexus at the global and basin levels", International Conference of Linkages Between Energy and Water Management for Agriculture in Developing Countries, Hyderabad, India. http://www.iwmi.cgiar.org/EWMA/files/papers/Energyprice_GW.pdf.

Annex 3.A1

Analytical model

Consider J wells pumping water for agricultural production from an aquifer during a given time interval. For each well j, pumping a quantity u_j produces net benefits in one year given by $B_j(u_j, \Theta_j, \Gamma)$ where $\partial B_j/\partial u_j \geq 0$ and $\partial^2 B_j/\partial(u_j)^2 < 0$ for all j. Concavity of the benefit function is a standard assumption. The vector Θ_j represents well-specific parameters that enter producer j's benefit function, such as the soil type, irrigation technology, field-level crop management choices and so on. The vector Γ incorporates parameters that influence the benefit function for all producers, such as input and output prices and weather. A crop production function is incorporated into the benefit function B_j, so that we can additionally define a pumping quantity $\overline{u_j}$ such that $\partial B_j/\partial u_j(\overline{u_j}, \Theta_j, \Gamma) = 0$ i.e. the marginal benefit of applied water becomes zero if enough water is applied. This means that even if water is available without cost, producer j will never apply more than $\overline{u_j}$. For simplicity, we assume that individual well benefit functions are independent, so that pumping at one well will not affect pumping at other wells. The general groundwater management problem is then given by

$$\max \sum_{j=1}^{J} B_j(u_j, \Theta_j, \Gamma) \text{ s.t. } \Phi(u_1, u_2, \dots, u_{J-1}, u_J, \Omega) \leq \overline{\Phi}$$

where $\Phi(u_1, u_2, \dots, u_{J-1}, u_J, \Omega)$ is a transfer function relating the pumping at all wells to the general hydrologic constraint $\overline{\Phi}$. The hydrologic constraint $\overline{\Phi}$ is a general function that describes the socially-desirable hydrologic outcome, in terms of aquifer, surface water, or groundwater-dependent ecosystem conditions. The benefit and transfer functions enter the maximisation through the construction of the Lagrangian function. The vector Ω represents all relevant parameters in the surface water-groundwater system, including hydrologic properties such as transmissivity and conductivity, and instream flow, other ecological, and legal constraints that must be met.

Case 1: Static model

For the simplest case, where all pumping has an equal impact on the aquifer in terms of saturated thickness, the transfer function Φ simplifies to the sum of pumping at all wells and $\overline{\Phi}$ is the desired aggregate pumping in a given year. The management problem may then be stated as

$$\max \sum_{j=1}^{J} B_j(u_j, \Theta_j, \Gamma) \text{ s.t. } \sum_{j=1}^{J} u_j \leq \overline{\Phi}$$

For this simple model, it is straightforward to show that the first-order conditions for the problem imply that $\partial B_i(u_i, \Theta_i, \Gamma)/\partial u_i = \partial B_j(u_j, \Theta_j, \Gamma)/\partial u_j \ \forall i, j$. In words, this means that an optimal allocation of water across all constrained groundwater users is one that equates their marginal benefits at each point in time. Note that in the absence of a pumping constraint, this model would predict that each producer would pump until the value of the marginal product of water were equal to zero, namely $\partial B_j(u_j, \Theta_j, \Gamma)/\partial u_j = 0 \ \forall j$. As the pumping constraint becomes more binding, the value of the marginal product of water will increase as $\partial B_j(u_j, \Theta_j, \Gamma)/\partial u_j \geq 0$. The optimal choice of marginal benefit depends on the degree to which pumping must be constrained to meet the aggregate pumping target $\bar{\Phi}$: the lower the desired value of $\bar{\Phi}$, the higher the marginal benefit needed to achieve it, as each producer must be constrained more. This result follows from concavity of the individual benefit functions.

If the hydrologic constraints are more complicated, for example if the aquifer properties vary across space, or if it is desirable to limit pumping further in some localised zones such as around drinking water supplies, then the general hydrologic constraint may be altered so that it incorporates the required spatial heterogeneity. So long as the impacts of pumping are effectively independent, the optimality conditions then equate the ratio of marginal benefit to the marginal impact on the hydrologic constraint across all users, namely:

$$\frac{\partial B_j(u_j,\Theta_j,\Gamma)/\partial u_j}{\partial\Phi(u_1,u_2,\dots,u_{J-1},u_J,\Omega)/\partial u_j}=\frac{\partial B_i(u_i,\Theta_i,\Gamma)/\partial u_i}{\partial\Phi(u_1,u_2,\dots,u_{J-1},u_J,\Omega)/\partial u_i}\ \forall i,j$$

The assumption of independence of pumping impacts corresponds to an application of the principle of superposition as used in analytical hydrologic modelling (e.g. Domenico, 1972; Freeze and Cherry, 1979). As such, this is generally not a strong assumption. Once again, the interpretation of the modified optimality conditions is straightforward. For a well that has a relatively higher impact on the aquifer, the marginal benefit at the optimal allocation will likewise be higher in order to satisfy the optimality condition, i.e. less pumping will be allowed.

Case 2: Simple dynamic model

We will extend the static model presented in Case 1 to allow consideration of externalities that may be lagged, i.e. that may produce impacts over multiple periods of time. Consider *J* wells pumping water for agricultural production from an aquifer during *N* separate increments of time. For each well *j*, pumping a quantity u_j^n at time $n=0,\dots,N$ produces net benefits given by $B_j(u_j^0,u_j^1,\dots,u_j^N,\Theta_j,\Gamma)$ where $\partial B_j/\partial u_j^n\geq 0$ and $\partial^2 B_j/\partial(u_j^n)^2<0$ for all *n*. The definitions of the vectors Θ_j and Γ are as before. Similarly, we assume that individual well benefit functions are independent, so that pumping at one well will not significantly affect pumping at other wells.

We assume that the basic goal of groundwater management is to choose a set of pumping paths to address issues of either aquifer depletion or surface water-groundwater interaction (sometimes referred to as stream depletion). The general management problem is then given by

$$\max\sum_{j=1}^{J}B_j(u_j^0,u_j^1,\dots,u_j^N,\Theta_j,\Gamma)\ \text{s.t.}\ \Phi(u_1^0,u_2^0,\dots,u_{J-1}^t,u_J^t,\Omega)\leq\bar{\Phi}(t)\ \forall t$$

where $\Phi(u_1^0,u_2^0,\dots,u_{J-1}^t,u_J^t,\Omega)$ is a transfer function relating the full pumping path at all wells to the relevant pumping constraint $\bar{\Phi}(t)$ at time $t\leq N$. The definition of Ω as representing relevant aquifer characteristics is as before. Note that discounting of future benefits may be incorporated into the benefits from pumping, $B_j(u_j^0,u_j^1,\dots,u_j^N,\Theta_j,\Gamma)$, without problem.

Externalities due to stream depletion caused by groundwater pumping may lead to the need for management if there are downstream or transboundary legal obligations related to surface water or if there instream impacts on habitat. In the former case of downstream or transboundary surface water obligations, the intent of regulation is generally to reduce pumping in order to limit cumulative stream depletion over a fixed interval such as a year or multiple years. In the latter case of instream habitat, regulations are intended to maintain minimum streamflow requirements throughout the year. In the model presented here, we will consider the design of policies to address cumulative stream depletion problems in some detail (Kuwayama and Brozović, 2013). Extension of the model to allow maintenance of specific streamflow constraints continuously throughout the year requires an intra-seasonal crop water production function, which leads to much more complex analyses and is generally not analytically tractable (e.g. Han, 2011).

A key feature of surface water-groundwater interaction is that stream depletion is a spatial and dynamic process and that because groundwater is a diffusional system, it is also subject to lagged effects (Glover and Balmer 1954, Sophocleous 2002). Thus, the impact of ongoing pumping on streamflow needs to consider the pumping history rather than just pumping in the current period. A general equation for stream impact from groundwater pumping at any time *T* after the start of pumping, for a well at a distance *d* from a stream, is then

$$\sum_{j=1}^{J}\sum_{s=0}^{T}\sum_{n=0}^{S} u_j^n \phi(d_j, T-n, \Omega_j) = \Phi(T)$$

In the equation above, Ω_j is the subset of hydrologic parameters relevant to pumping at well *j*. By assumption, the stream depletion externality is linear in pumping (this is also an application of the principle of superposition in hydrology; Domenico, 1972; Freeze and Cherry, 1979). Because of this, the transfer function ϕ may be interpreted as the marginal externality of pumping at time *n* occurring at time $T > n$. Thus, the equation represents the sum of lagged impacts occurring *at time T* from all pumping that occurred at or before time *T*.

Hydrologic stream response functions can be used to model the exact relationship between pumping and stream flow, accounting for the fact that significant time lags exist between pumping decisions and the consequent stream depletion, and that the magnitudes of these time lags depend primarily on the distance, *d*, between wells and nearby streams. In addition to their use for designing groundwater regulations, analytical methods are also widely used by practitioners for general assessments of stream depletion.

Where detailed numerical groundwater models (e.g. MODFLOW) are available, these have been used to determine the impact of pumping on stream depletion. In Nebraska, numerical methods have been used in the Republican River Basin and the Big Blue River Basin (MODFLOW-based), and in the Platte River Basin (COHYST-based). Elsewhere, analytical and graphical methods based on solutions of the transient groundwater flow problem applied to the case of surface water-groundwater interaction are applied. Hydrologists have derived stream response functions for use in a variety of different hydrologic settings.

The analytical solution by Glover and Balmer (Glover and Balmer, 1954) is one of the simplest analytical solutions but because it has been widely applied in a policy context, we will discuss it here (Glover and Balmer, 1954, Jenkins, 1968; Nebraska DNR, 2007). For this case, assume that $\Omega_j = \Omega_j(S, \tau)$ where *S* is the aquifer storage coefficient and τ is the aquifer transmissivity (units are square feet per year). Stream depletion caused by well *j* after *t* years of pumping at a constant rate, measured in acre-feet per year (1 acre foot is equal to 1.223 megalitres), is given by the Glover-Balmer equation as

$$\phi(d_j, t, \Omega_j) = u_j \operatorname{erfc}\left(\sqrt{\frac{d_j^2 S}{4\tau t}}\right)$$

where *d* is the distance between well and stream, *t* is the time in years since the start of pumping, and erfc is the complementary error function. This equation can be modified to account for seasonal pumping. Similarly, other more sophisticated versions are available for surface-water groundwater interactions such as partially penetrating wells or streambed clogging (e.g. Hunt, 1999; Hunt, 2012).

Given the management problem above and the Glover-Balmer equation, the first-order conditions for the problem can be used to show that $\partial B_j(u_j^0, u_j^1, \dots, u_j^N, \Theta_j, \Gamma)/\partial u_j^i - \lambda \sum_{s=i}^{T} \phi(d_j, T-s, \Omega_j) = 0 \; \forall i, j$, where λ is the Lagrange multiplier. It follows that for an interior optimum:

$$\frac{\partial B_j(u_j^0, u_j^1, \dots, u_j^N, \Theta_j, \Gamma)/\partial u_j^i}{\sum_{s=i}^{T} \phi(d_j, T-s, \Omega_j)} = \frac{\partial B_l(u_l^0, u_l^1, \dots, u_l^N, \Theta_l, \Gamma)/\partial u_l^k}{\sum_{s=k}^{T} \phi(d_l, T-s, \Omega_l)} \quad \forall i, j, k, l$$

Thus, the ratio of the marginal benefit from pumping to the marginal externality caused by pumping should be equal across all well locations. The Lagrange multiplier λ may then be interpreted as the effective (present value) entitlement price. If the marginal damage of the externality is equivalent for all firms, this outcome can be induced with marketable entitlements that are traded on a one-to-one basis, where marginal abatement costs of all firms will equal marginal damage multiplied by λ (Kuwayama and Brozović, 2013; Palazzo and Brozović, 2014). Conversely, if the marginal benefit function is the same at each pumping location, so that $\partial B_j(u_j^0, u_j^1, \dots, u_j^N, \Theta_j, \Gamma)/\partial u_j^i = \partial B_l(u_l^0, u_l^1, \dots, u_l^N, \Theta_l, \Gamma)/\partial u_l^k = B'(u^0, u^1, \dots, u^N, \Theta, \Gamma)$, then wells closer to the stream will always be more constrained than wells further from the stream i.e. $u_j^i < u_l^i$ for all *i* and $T > i$ if $d_j < d_l$ (Kuwayama and Brozović, 2013). To show the latter result, consider two wells *j* and *l* with $d_j < d_l$. Then it must be that $\sum_{s=i}^{T} \phi(d_j, T-s, \Omega_j) > \sum_{s=i}^{T} \phi(d_l, T-s, \Omega_l)$ for all *i* and $T > i$ (e.g. consider the Glover-Balmer equation). The result follows immediately from the optimality conditions, as it implies that $B'(u_j^0, u_j^1, \dots, u_j^N, \Theta, \Gamma) > B'(u_l^0, u_l^1, \dots, u_l^N, \Theta, \Gamma)$, and because $B'' < 0$, it must be that $u_j^i < u_l^i$.

Annex 3.A2

Case study: Choice of policy instruments for groundwater management

The most efficient way to achieve any required environmental outcome is with an appropriately targeted incentive-based scheme. However, the gains from optimal regulation relative to other, simpler kinds of regulation may not be quantitatively important and depend both on the heterogeneity of producers and their production functions, and on the spatial complexity of the hydrologic system. For any given application, it is important to understand whether the gains from the introduction of a complex regulatory or incentive-based system for groundwater management are worthwhile, in comparison to alternate regulations with lower informational requirements and higher stakeholder acceptability.

Annex 3B.1 showcases a comparison of alternate policies to reach stream flow goals in an agricultural watershed with widespread groundwater pumping for irrigation (Brozović and Young, 2014). The case study area is the Upper Republican Natural Resources District (URNRD) in Nebraska (Figure 3.2).[1] This groundwater management district overlies the High Plains aquifer and must reduce aggregate groundwater pumping to meet stream depletion goals as a result of interstate litigation (McKusick, 2002). All 3 200 wells in the district are metered and groundwater pumping entitlements at each well are quantified and enforced. Pumping is only allowed on certified irrigated acres and there is a moratorium on new wells or acres, so the maximum irrigated land area is fixed.

In the model, profit functions are calibrated for each well that consider the joint land use, crop choice, and applied water decision (Martin et al., 2007; Palazzo and Brozović, 2014). The model feeds on several types of spatial data at the well level, including information on acreage irrigated by each well, depth to water and well yield, soil type, crop evapotranspiration requirements, and irrigated and dryland crop yields for corn, wheat, soybeans, and sorghum (the major crop types in the study area). For each set of well-specific parameters, a nonlinear optimisation is employed to determine crop choice, land use, water application, and expected profits (Palazzo and Brozović, 2014). The baseline water availability for the analysis is the current regulation for the URNRD, 13 acre-inches per year for each certified irrigated acre. Available water for each well is then sequentially reduced in order to estimate the marginal benefits of water use in irrigation and the resulting foregone profits. Adjustment to reductions in water availability is allowed at both the extensive and intensive margins (English, 1990; Palazzo and Brozović, 2014). Next, the set of profit functions can be used to compare the trade-offs between aggregate water use reductions, resulting improvements in stream flow, and producers' foregone profits.

The analysis uses the Stream Depletion Factor (SDF) to estimate the stream depletion externality. Following URNRD rules, the SDF is defined as the proportion of water pumped from a well that is drawn from an adjacent stream. SDFs used in the analysis are calculated over a 50-year time horizon with seasonal pumping for irrigation and the Glover-Balmer equation parameterised with hydrologic data for the Republican River Basin (Kuwayama and Brozović, 2013). As stream depletion is modelled

1. Local administration areas are responsible for developing and implementing groundwater management policy and are called Natural Resources Districts (NRDs) in Nebraska, Groundwater Management Districts (GMDs) in Kansas, and Groundwater Conservation Districts (GCDs) in Texas.

as an additive process, the SDF is also equal to the marginal externality of an additional unit of pumping.

Figure 3.B1. High Plains Aquifer region and selected groundwater administration areas

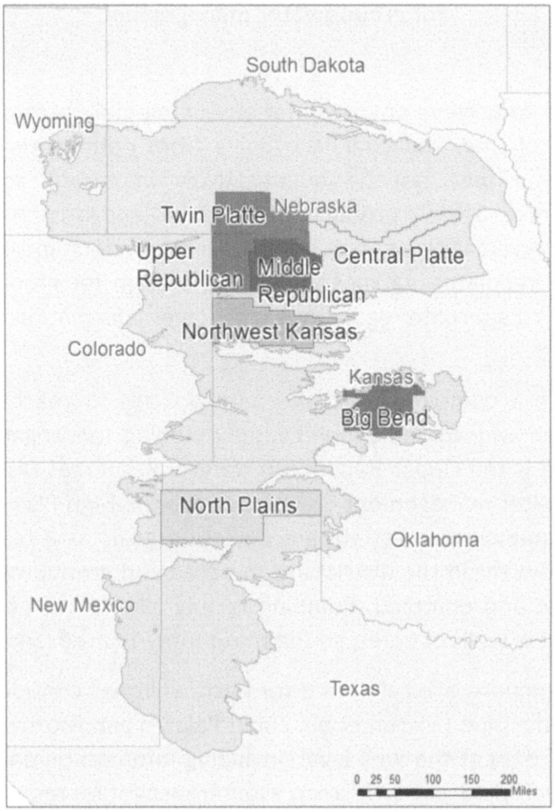

In particular, three different kinds of policies are analysed, each of which has been implemented in the watershed of interest: pumping quotas, irrigated land retirement, and transferable groundwater pumping entitlements. In each case, the current water allocation in the URNRD provides the baseline for measurement. For each of the alternate policies, aggregate water use is reduced by varying amounts and then compared to the total foregone profits required to attain that water use. For the pumping quotas, it is assumed that they are equally applied to all producers. For the land retirement program, three types of targeting options are considered: (i) based on the cheapest land (reducing irrigated acreage fastest); (ii) based on the cheapest water (reducing aggregate pumping fastest); and (iii) based on the highest stream depletion (shutting down wells with the highest marginal externality first). For the transferable entitlement system, frictionless trading is assumed, noting that metering, quantified allocations and enforcement are already in place in the URNRD.

Lastly, two types of transfers are considered: those that are unadjusted for stream depletion where the unit of transfer is quantity of water (here called "simple") and first-best transfers that are adjusted for stream depletion, where the unit of transfer is quantity of stream depletion (here called 'complex'). For the simple transfer scheme, marginal benefits of pumping water are equalised across all wells without any adjustment for differences in the spatial stream depletion externality. This corresponds to trading across the district at a single market price, equal to the marginal benefit. For the complex transfer scheme, the marginal benefits at each well are normalised by the expected impact on stream depletion (as described in the theoretical development above). In this case, each well faces an idiosyncratic price for pumped water, corresponding to a single market price for expected stream depletion resulting from pumping. In all cases, estimated SDFs are used to quantify the stream depletion externality (Kuwayama and Brozović, 2013).

Results

Figure 3.B2 shows the results expressed in terms of annual costs. By definition, when stream depletion reduction is the policy target, a complex transferable entitlement scheme that adjusts for stream depletion will be the cost-effective method of achieving any instream target (Figure 3.3). Perhaps surprisingly, the simple transferable entitlement scheme is also cost-effective for small reductions in stream depletion. The simple entitlement scheme does not adjust for stream depletion, but in the URNRD there are a number of wells with high stream depletion impacts but very low value of the marginal product of water. These wells will be sellers in any groundwater market, whether there is an adjustment for stream depletion or not.

Figure 3.B2. Comparison of the cost-effectiveness of alternative policies for reducing stream depletion impacts in the URNRD, Nebraska

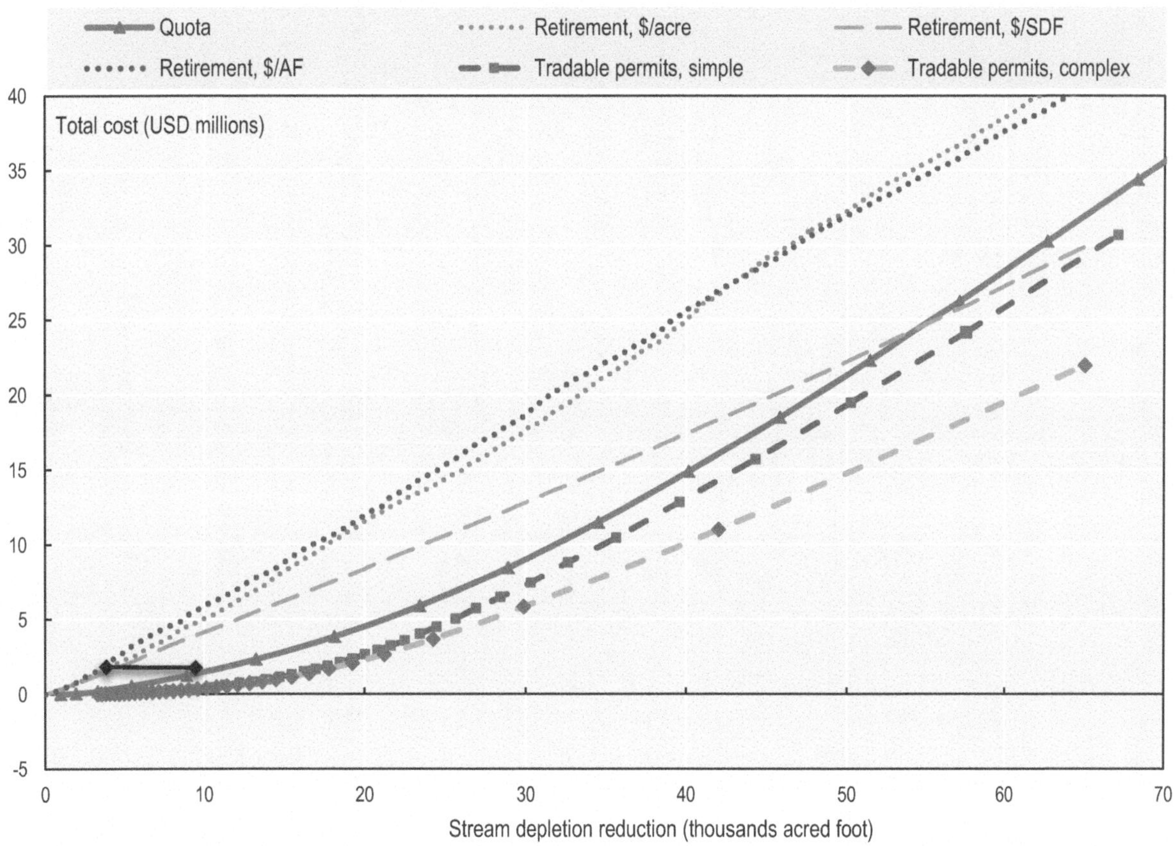

Note: SDF is the stream depletion factor. AF: acre foot.
The two dots joined by a line represent the estimated annualised cost of the UNRNRD's Rock Creek project, with costs and depletion reductions shown for irrigated acreage retirement (left dot) and retirement together with stream augmentation pumping (right dot).
Source: Brozović and Young (2014).

The uniform quota is also more cost-effective than the land retirement schemes until large reductions in stream depletion are needed, when the land retirement policy targeted on stream depletion performs better. Note that the URNRD's implemented Rock Creek land retirement and stream augmentation project is similar in cost to the predicted costs of land retirement (left hand dot on the horizontal line in Figure 3.3). With stream augmentation in operation, but without considering the energy costs of pumping, the project is similar to a reduced quota in cost-effectiveness.

Chapter 4

What agricultural groundwater policies exist in OECD countries?

This chapter reviews policies and management approaches for agricultural groundwater management in OECD countries. Responses to a questionnaire are used to examine the diversity of national and regional policy instruments. An analysis is also conducted at the regional level to assess whether the choice of management instruments can be linked to specific characteristics and constraints of agricultural groundwater systems.

The statistical data for Israel are supplied by and under the responsibility of the relevant Israeli authorities. The use of such data by the OECD is without prejudice to the status of the Golan Heights, East Jerusalem and Israeli settlements in the West Bank under the terms of international law.

Key messages

There is a wide diversity of policies and approaches applied to manage groundwater use for agriculture in OECD countries. Results from a country survey sent to the 34 OECD member countries, which elicited 20 usable responses, show wide variations. Groundwater policies are founded on different legal systems; focus is on demand-side and/or supply-side; and direct and indirect approaches to regulatory, economic or collective management are used.

There are differences at the country, regional and sub-regional levels, and few broad similarities can be identified across countries.

Regulatory approaches are the most common management approach, serving as a basis for other instruments, but specificities across entitlement regimes also vary widely; from allocation mechanisms, to their nature and scope.

At the regional level, there is no evidence of a clear relationship between the overall scope of management approaches (defined as the number of approaches to control groundwater use) and the intensity of groundwater stress (based on agro-climatic pressures and externalities); instead, clusters of regions share similarities. Some OECD regions facing relatively high groundwater stress use a comprehensive range of groundwater management instruments (economic, regulatory, direct, indirect, etc.). But other regions facing equally high degrees of stress use much fewer instruments; and some of the regions facing the least stress apply a relatively wide range of instruments.

The analysis also shows that some management instruments are partially correlated with specific constraints. Economic and supply-side approaches are more prevalent in areas with higher agricultural groundwater stress.

How should these policies evolve to help improve agriculture groundwater management? Responding more effectively to high groundwater stress and associated externalities observed in OECD regions may require policy reform. Chapter 5 will use findings from this chapter and the conclusions of Chapter 3 to identify policy recommendations to help move towards a more sustainable groundwater management model.

An analysis based on findings from a 2014 OECD survey on groundwater management approaches

This chapter reviews governmental policies in OECD countries supporting sustainable groundwater management for agriculture. While a number of these countries use groundwater in agriculture, others are well endowed and not using it much, or face the reverse issue of underuse of groundwater resources. Even within the first group, there is a large range of regional settings, reflecting the degree of use of groundwater resource, whether there is overdraft, and the presence and intensity of externalities (Chapter 2). The chapter will analyse whether this diversity of constraints is matched by a set of corresponding policies.

A number of articles, book chapters and reports have been published on groundwater policies and agriculture, but they tend to focus on specific countries or regions of interests (e.g. Grafton et al. 2014; Shah, 2008; de Stefano and Llamas, 2012). Gardũno and Foster (2010) draw transversal lessons from experiences in multiple groundwater irrigating regions, but they focused on developing countries mostly outside of OECD. Morris et al. (2003)'s international study approached groundwater use from a much broader perspective with limited agricultural perspectives. And several publications either compare policies in two countries or regions (e.g. Scott and Shah, 2004), draw learnings from one country for wider settings (Garrido et al., 2006), or combine collections of chapters about experience in various countries (e.g. Giordano and Villholth, 2007). This chapter intends to complement this literature by proposing an OECD wide multi-country transversal comparison of policy approaches.

This chapter is primarily based on the analysis of responses to an OECD questionnaire launched in the summer of 2014, assessing the status of resources and agriculture use and listing the relevant policies at the national and regional level. The questionnaire was structured into three parts, with the first providing general information on the respondent. The second focused on characterising the status and use of groundwater resource in agriculture at the national and sub-national level, mostly to feed into Chapter 1. Respondents were asked to select one to four groundwater regional units (henceforth called regions) for which they could provide more information about groundwater use and the constraints thereof. The last part focused on the presence of a wide range of instruments (Table 3.1) that could potentially affect groundwater. The full questionnaire is available upon request.[1]

The questionnaire was sent to delegates of the 34 OECD countries and the European Union. As shown in Table 4.1, 27 countries submitted at least partial responses. Of these, seven countries did not provide sufficient usable information about policies (either doe to the absence of specific policies or potentially partial access to information), leaving a set of 20 OECD countries with usable responses. Thirteen of these countries provided at least some information on 27 agriculture groundwater regions.

Survey responses are complemented by information from a comprehensive review of the literature on groundwater management in agriculture, drawing on examples of individual regions using groundwater for agriculture. Naturally, each of these regions presents specific agricultural groundwater system characteristics and therefore may not provide ubiquitously replicable management solutions.

Table 4.1. Coverage of received responses to the OECD groundwater questionnaire[1]

Country	Region	Policy responses
Australia	Great Artesian Basin, Murray-Darling Basin	Limited
Austria		
Canada		No
Chile		
Czech Republic		
Denmark	Western Jutland	
Estonia		
Finland		
France	Nappe de la Beauce and Département de la Vienne	
Greece		No
Ireland		No
Israel	Western Galilee	Partial
Italy	Puglia, Campania (Ufita)	Limited
Japan	Kinugawa Seibu; Noubiheiya Seibu; and Kikuchi Heiya	
Korea	Jeju volcanic island	
Mexico	Region Lagunera	
Netherlands	Meuse River (North-Brabant), Sand Meuse River Basin (Limburg), Rhine-East River Basin (Gelderland), and Rhine-East River Basin (Overijssel).	
Poland		No
Portugal	River Basin District of Tejo e Ribeiras do Oeste	
Slovak Republic		
Slovenia		No
Spain	Mancha Occidental; Campos de Montiel; Almonte-Maritias; and Mancha Oriental.	
Sweden		
Switzerland		No
Turkey	Küçükmenderes Basin	Limited
United Kingdom		No
United States	Northern High Plains Aquifer (NHPA); Southern High Plains Aquifer (SHPA); Mississippi Alluvial Aquifer (MAA) Region; and Mountain and Pacific West (MPW) Region	

1. 2014 OECD questionnaire on groundwater use in agriculture.

The next section reviews agriculture groundwater policy approaches in OECD countries, using the instrument categories outlined in Table 3.1. The following section explores more specifically whether the choice of policies coincides with the challenges faced by agricultural groundwater management systems using responses at the regional level.

A wide spectrum of agricultural groundwater management approaches

Demand-side management approaches: Shared core approaches, diverse instruments

Different legal status and entitlement characteristics stemming mostly from legal traditions

Groundwater is generally considered under water laws as part of broader legislation on water (Mechlem, 2012). The approaches may be piecemeal, addressing quality or quantity concerns, or more comprehensive but there is an increasing effort to combine all surface and groundwater legislation (Mechlem, 2012).

Nineteen of the 20 countries responding to the questionnaire reported national groundwater reforms, including 15 during the last ten years. EU member states have transcribed in their own legislation the EU Water Framework Directive of 2000 (EC, 2000), requiring quantitative management of groundwater resources, and the additional EU Groundwater Directive (EC, 2006) which focusses on quality concerns. As explained below, these two laws outline comprehensive groundwater management plans at the regional level.

Reform of water allocation can be triggered by different factors, including water scarcity and ecosystem risks (OECD, 2015c). Similarly, specific changes in policies pertaining to groundwater can be triggered by crises and/or conflicts. Two main concerns underlie recent changes in groundwater policies in some OECD countries: long-term aquifer depletion and surface water-groundwater interaction (e.g. McCarl et al., 1999; Scanlon et al., 2012). In the United States, concerns about groundwater pumping externalities are manifested in ongoing litigation over water resources and rapidly changing water management institutions (Hathaway, 2011; McKusick, 2002). Surface water-groundwater interaction has also been a major driver for changes in groundwater policy (Kuwayama and Brozović, 2013; Palazzo and Brozović, 2014); surface water flows are the subject of both transboundary legal challenges over river basin allocations and potential environmental impacts to instream habitat and other groundwater-dependent ecosystems (e.g. McKusick, 2002; Delaware River Basin Commission, 2008).

Groundwater use is generally defined based on the use of entitlements (licenses, permits or rights), generically defined in this report as the permission to abstract and use groundwater from an aquifer system as specified under the relevant legal texts (see the glossary at the end of the report). These entitlements are the cornerstone of most regulatory approaches for groundwater management, "the central element of groundwater laws" (Mechlem, 2012), and have been found to be critical in reducing groundwater overdraft (Kemper, 2007). Groundwater can have public or private ownership, and the status of ownership does affect how entitlements are allocated and potentially used.

The legal status of groundwater ownership varies by country or region within the OECD area. Unlike in the case of surface water, groundwater traditionally remains in the private domain in multiple countries (OECD, 2010a and 2015c). Of the 22 national or regional responses listed in the first column in Table 4.2,[2] 12 have at least partly private ownership of groundwater.[3] Groundwater ownership is also typically linked to land ownership, unlike surface water which is often unrelated to land and overwhelmingly held publically (e.g. 88% public ownership reported in OECD, 2015c). The characteristics of groundwater as locally defined and related to specific land have made it legally appropriable by private actors, even if groundwater bodies often comprise a collective of users.

Table 4.2. Groundwater entitlement characteristics by responding regions or countries

Ownership	Entitlement duration and characteristics	Beneficiaries of entitlements	Doctrines used as a basis for allocation
Private Austria, Japan, Portugal; MAA and MPW regions	**Permanent** Chile, France, Korea, Slovak Republic, Sweden, Turkey, United Kingdom; Murray-Darling Basin (Australia), NHPA, SHPA, MAA and MPW regions	**Individuals** Austria, Chile, Czech Republic, Finland, France, Israel, Japan, Korea, Portugal, Mexico, Sweden; Mancha Occidental, Campos de Montiel, and Almonte-Marismas and Mancha Oriental (Spain), Murray-Darling Basin (Australia), NHPA, SHPA, MAA and MPW regions	**Absolute ownership** Chile, MPW region
Public Estonia, France, Netherlands, Spain; Murray-Darling Basin (Australia)	**Temporary** Austria, Chile, Czech Republic, Estonia, Israel, Korea, Mexico, Netherlands, Portugal, Slovak Republic, Spain, Sweden, Turkey, United Kingdom; Nappe de Beauce (France)	**Collective bodies** Austria, Chile, Czech Republic, Finland, Israel, Mexico, Portugal, Sweden; Mancha Occidental, Campos de Montiel, Almonte-Marismas (ESP), Murray-Darling Basin (AUS)	**Reasonable use** Estonia, Finland, France, Korea, Mexico, Portugal, Spain, Sweden; SHPA, MAA and MPW regions
Both Chile, Denmark, Korea, Mexico; Sweden; NHPA and SHPA regions	**Linked to land rights** Finland, Israel, Japan, Korea, Sweden, United Kingdom; Departement de la Vienne and Nappe de Beauce (France), Murray-Darling Basin (Australia); NHPA, SHPA, MAA and MPW regions	**Companies** Austria, Chile, Czech Republic, Finland, Israel, Mexico, Portugal, Sweden; Almonte-Marismas (Spain), Murray-Darling Basin (Australia), NHPA, SHPA, MAA and MPW regions	**Correlative rights** Chile, Estonia, Finland, France, Israel; Murray-Darling Basin (Australia), NHPA and MPW regions
Other or neither Canada, Czech Republic, Finland, Slovak Republic, Turkey	**Transferable** Chile, Korea, Mexico, Spain; Murray-Darling Basin (Australia); NHPA and SHPA regions;	**Other** Chile, Finland, Slovak Republic	**Prior appropriation:** Chile, France, Sweden, SHPA and MPW regions

Note: NHPA: Northern High Plains Aquifer, Untied States; SHPA: Southern High Plains Aquifer, United States; MAA: Mississippi Alluvial Aquifer, Untied States; MPW: Mountain and Pacific West, Untied States; UK: United Kingdom.
Source: 2014 OECD questionnaire on groundwater use in agriculture.

Groundwater entitlements may, on the other hand, share similar characteristics with those employed for surface water, even if they remain less employed. Twelve responding regions or countries define permanent entitlements, while 13 use temporary entitlements. The renewal time often depends on the type of water use (OECD, 2015c). In particular, some countries tend to have periodic limits specifically for abstraction for irrigation purposes (e.g. 12 years for Austria), while other limits are applied to other uses. In seven responses, it was noted that these rights are transferable, which can imply the possibility of permit ownership transfers, for instance linked to a change in land ownership, or may open the door for potential transactions and or markets. The beneficiaries of groundwater entitlements include mostly individuals, with a number of countries and regions allowing companies and collective bodies.

Many of the presented groundwater entitlement systems are effectively linked with land property rights, which is not ubiquitous for surface water. Bundling groundwater with land property rights can make it more difficult for resource management as it leaves less freedom of operation for users (OECD, 2015c). At the same time, this historic linkage is found in multiple regions on other underground resources. Mechlem (2012) reports there is a worldwide trend towards separating groundwater rights from land rights and moving towards publically-owned resources for which users can operate entitlements. However, such reforms still appear to be outside the scope of a number of OECD countries.

Looking across these categories — first three columns of Table 4.2 — a few weak patterns may emerge, with a group of countries or regions with private individual entitlements, like Austria, Japan, the Mississippi Alluvial Aquifer (MAA) and Mountain and Pacific West (MPW) regions, and the United States, and another group with multiple status and characteristics, like Chile, Korea, Mexico, and Northern and Southern High Plains Aquifer (NHPA and SHPA). France, on the other hand, uses individual public permanent rights. These statuses and characteristics seem to stem more from legal traditions than from physical considerations of groundwater characteristics.

Beyond their nature, the allocation system of entitlements is critical in considering both the functioning and equity of groundwater management systems. Four main doctrines have been used in Western management systems (Joshi, 2005; Peck, 2007; Wichelns, 2010): absolute ownership (also known as "Rule of capture" or "English rule"); reasonable use; correlative rights; and prior appropriation (see Glossary for complete definitions). Multiple sub-categories can also be found in detail, combining the characteristics of the main doctrines. Table 4.2 shows that, once again, a diversity of approaches is used in OECD countries. Box 4.1 illustrates the diversity of doctrines used in the United States.

Each of these doctrines has advantages and drawbacks, in terms of degree of freedom for owners and administrative costs to implement (Peck, 2007). The absolute ownership doctrine is simpler to use and presents the lowest public costs for water management, but can lead to conflicts and insecure resources (Joshi, 2005). It also can lower the cost of access to the resource compared to surface water, which may in some cases lead to stream depletion (e.g. OECD, 2012b). Reasonable use encourages accounting for harm done to neighbours, but depends on the specific terms and interpretation of what is reasonable. Correlative rights enable more regional management and prior appropriation, while raising equity concerns enables regulating wells.

The linkages between water ownership, water entitlements, and land property rights are important to understand the scope and limits of public policies in agricultural groundwater use. A large share of land is used for agriculture in OECD countries. If the right to use groundwater is directly associated to land, then, mathematically, agriculture is bound to be better allocated than other water using sectors. But it also implies that groundwater resources will likely have an impact on land values and land use. As developed below, this also explains why a number of groundwater instruments target land rather than groundwater or agriculture.

Box 4.1. Groundwater allocation in the United States: A patchwork of systems

Groundwater rights' systems are more complex than those involving surface water in the United States. States set their own groundwater laws, supporting diverse ways to manage groundwater. In particular, doctrines have been used for groundwater allocation. Figure 4.1 shows an assessment of the repartition of doctrines by State, as of 2005, showing the importance of prior allocation in the West, and reasonable use in the East, with pockets of other States using Absolute ownership or mixed systems. Among the eight states underlying the High Plains Aquifer (TX, OK, NM, KS, CO, NE, SD, and WY), at least three doctrines can be found. These systems have evolved over time; Kansas, for instance, moved from absolute ownership to prior appropriation in 1945. All these systems are complemented by federal laws.

While the effectiveness of the approaches can be compared, these differences can raise tensions across states for the management of interstate groundwater bodies and indirectly for surface water, especially in periods of droughts.

Figure 4.1. Simplified map of groundwater allocation systems in the United States, as of 2005

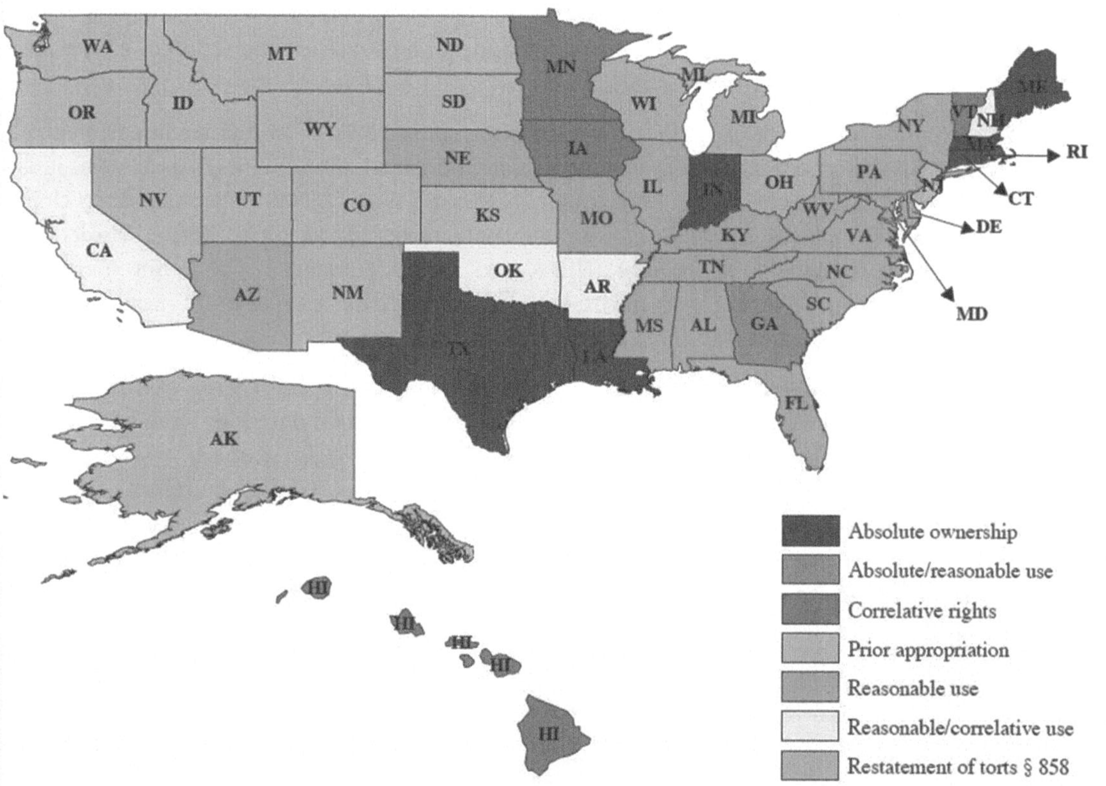

Source: Joshi (2005), http://aquadoc.typepad.com/files/gw_rights_thesis.pdf; Joshi (2005), Wichelns (2010), Peck (2007) and Sophocleous (2010).

Widely used management plans and groundwater regulations, often operating at the sub-national level, face enforcement challenges

Thirteen of the 20 responding countries, and 14 of the 20 groundwater regions they included, report having groundwater management plans. Most of the plans are mandatory at the national level, whereas a large share is voluntary at the regional levels. In the United States, groundwater is managed at the state level and often implemented in groundwater management districts, with varied legal requirements and voluntary schemes (Wichelns, 2010). The EU Water Framework Directive requires setting management units, plans for actions and monitoring systems to reach good quantitative status for all defined bodies by 2015 (Box 4.2). Two responding countries report having no plans: Chile and Japan (at the national or regional level).

Box 4.2. Managing groundwater at the sub- river basin level: The 2000 EU Water Framework Directive

The EU Water Framework Directive (WFD) addresses multiple quantitative and qualitative objectives for surface and groundwater. The core elements of the groundwater components of the Directive relate to the establishment of groundwater bodies under river basin districts, for which groundwater used will be monitored and regulated to achieve what is defined as good quantitative and chemical status by 2015. More specifically it requires EU countries to:

- Define groundwater bodies within each of the national river basin districts.
- Establish registers of protected areas within each district.
- Establish groundwater monitoring networks to assess the status and evolution of groundwater bodies towards good quantitative and chemical status.
- Set up river basin management plans (RBMPs), which report the status of groundwater and account for pressures on groundwater bodies, to be published in 2009 and 2015.
- Application of the principle of recovery costs for water services by 2010.
- Establish a program of measures to achieve environmental objectives by 2012, including for instance groundwater extraction controls, controls of artificial recharge.

As noted in Chapter 1, good quantitate status is defined as: "The level of groundwater in the groundwater body is such that the available groundwater resource is not exceeded by the long-term annual average rate of abstraction. Accordingly, the level of groundwater is not subject to anthropogenic alteration such as would result in: (a) failure to achieve the WFD environmental objectives for associated surface waters, (b) any significant diminution in the status of such waters, and (c) any significant damage to terrestrial ecosystems which depend directly on the groundwater body. Alterations to flow direction resulting from level changes may occur temporarily, or continuously in a spatially limited area, but such reversals do not cause saltwater or other intrusion, and do not indicate a sustained and clearly identified anthropogenically induced trend in flow direction likely to result in such intrusions." These objectives therefore include the diminution of overdraft and addressing externalities associated with groundwater use.

As of 2009, 11 897 groundwater bodies had been defined and assessed for the 19 OECD EU countries with official reports in 2009 (Austria, Belgium, Czech Republic, Denmark, Estonia, Finland, France, Germany, Hungary, Italy, Ireland, Luxemburg, Poland, the Netherlands, Slovak Republic, Slovenia, Spain, Sweden, and the United Kingdom). But there was still a significant share of bodies with unavailable information.

The additional Groundwater Directive of 2006, which was subject to revisions in 2014, focuses primarily on quality, with the objective of preventing the intrusion of pollutants and other dangerous substances. Other EU directives provide further quality requirements applying to groundwater (e.g. the Nitrates Directive and Plant Protection Products Directive which look specifically at agriculture-induced water quality concerns).

Source: EC(2000), EC (2006), EUWI Med (2007), http://ec.europa.eu/environment/water/water-framework/groundwater/framework.htm

Groundwater is under the responsibility of different types of institutions. Seven of the 20 responding countries point to Ministries of Environment and that of Natural Resources or their equivalent, another to the central government, one to a national water agency, and four to local or regional water institutions.[4] In most other countries, there are multiple national authorities in charge of one or another aspect of groundwater management; for instance, six in the United States (US Geological Survey, US Environmental Protection Agency, US Department of Agriculture, US Bureau of Reclamation, US Army Corp of Engineers, US Bureau of Land Management), three in Italy and Portugal, four in Korea. This may require challenging inter-agency co-ordination. At the same time, the management of groundwater is at least partially devolved to the regional level, even if the type of region or institution in charge also diverges widely in all responding countries (Figure 4.2). Over 30%

of the respondents point towards administrative regions, river basins, or groundwater management bodies. Some innovative mechanisms may also include co-responsibility for management, as observed in the case of the Eastern Mancha aquifer in Catalonia and the Duero Basin in Spain (Lopez-Gunn et al., 2012b).

Figure 4.2. Geographical levels of subnational groundwater management among responding countries

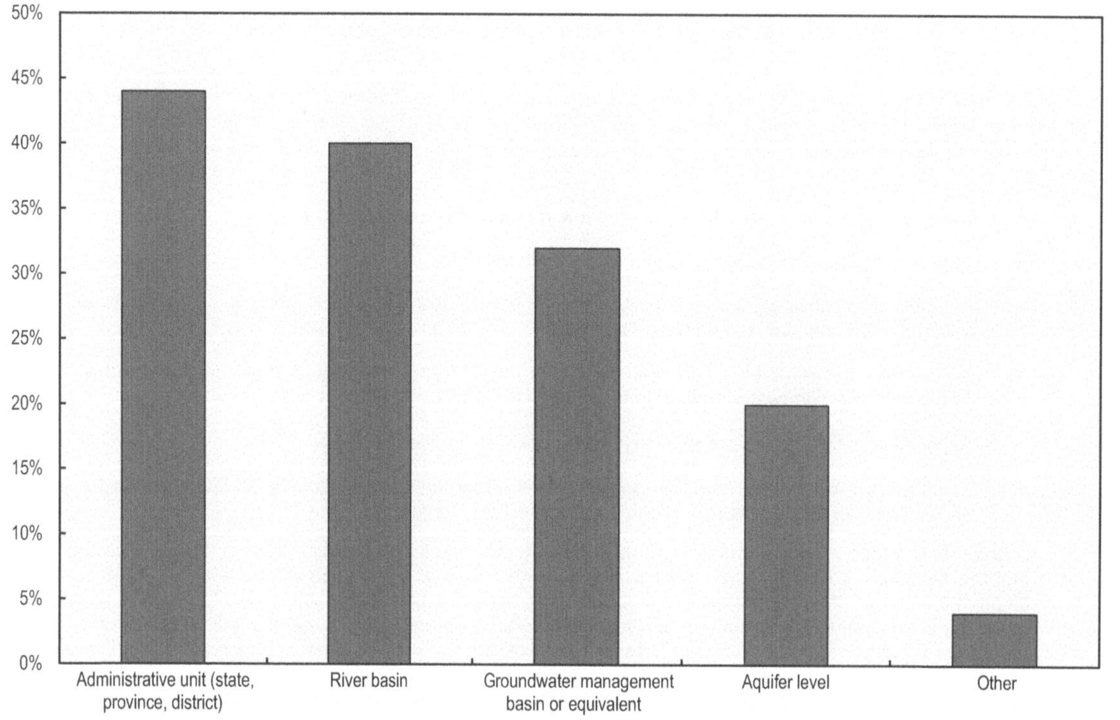

Source: 2014 OECD questionnaire on groundwater use in agriculture.

These management plans do not specifically cover agriculture. In a supplementary question, 56% of the responding countries reported that agriculture was a main, if not the major, user of groundwater in at least one of these regions. These countries include the leading OECD ten countries in groundwater use for irrigation (Chapter 1): Australia, Greece, Italy, Japan, Korea, Mexico, Portugal, and Spain. A large majority of national responses (80%) also note that groundwater management is at least partially linked with surface water. It is systematic for most countries or regions with mandatory management plans, but there are exceptions. Countries with no management plans can also have systematic connections, and some regions with mandatory plans only link the two in a limited fashion.

Separating surface and groundwater management is considered one of the main sources of problems in groundwater management (e.g. OECD, 2010a). Historically, water laws started to focus on surface water because groundwater was less visible and subject to pressure, but as some countries evolved, others lagged behind (Mechlem, 2012). The lack of willingness to connect surface and groundwater in policy decisions were observed particularly in Spain and can be attributed to four causes: the lack of human and technological capacity, limitations of water legislation, social and bureaucratic constraints and political factors (Llamas, 1975). This separation resulted in mismanagement of groundwater (Llamas, 2004). In their international political economy study on groundwater management in semi-arid countries, Garrido et al. (2006) considered this as the first stage in policy making, as had been the case for Spain, India, Mexico, and the US states of California, Texas and Arizona (Chapter 1).

Three regulations are generally considered in plans or legal frameworks: those pertaining to wells, metering, and withdrawals. In parallel, or when groundwater cannot be subject to such regulation, indirect regulatory instruments can also be used via restrictions on irrigated land. Figure 4.3 reports the modes of approaches used by responding countries and regions. Regulations on wells, via permits or authorisation are the most often used approaches, followed by quantitative restrictions (quotas) and metering requirements. Regulations on the expansion of irrigated lands come last. All these regulations are more prevalent at the regional than the national level. Indeed, five of the responding countries report that such regulations operate at the regional rather than national level (Australia, Denmark, Japan, Netherlands and the United States), which points to potential sub-national specificities in requirements.

Figure 4.3. Number of OECD countries or regions using specific groundwater regulations

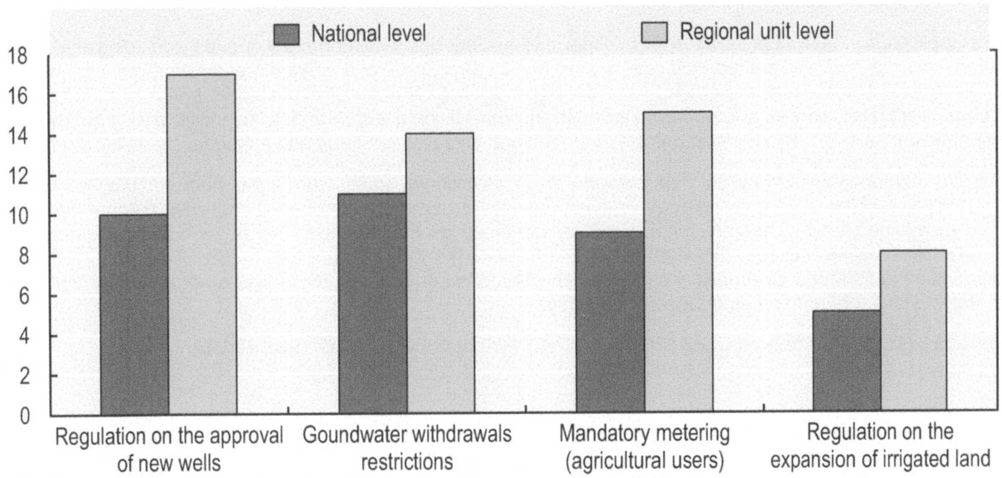

Source: 2014 OECD questionnaire on groundwater use in agriculture.

More specifically, regulations on the *approval of new wells* accounts for well spacing and environmental assessments in most responses (70% and 67% of combined responses, respectively). Well spacing can provide guarantees against well interference and stream depletion if implemented with sufficient knowledge of the situations. Its application may also vary depending on the local conditions and constraints; in the United States well spacing requirements vary from 300 feet (100m) or less in Texas to 4 miles (6.4 km) in portions of the Dakota aquifer in Kansas (Brozović et al., 2006). Environmental assessments can go beyond and assess potential externalities resulting from the new well and its uses to support or reject proposals, but as in other areas, their scope and methods are important as is the role of public participation and transparency. In Prince Edward Island, Canada, a political debate was launched in 2014 about the use of deep wells for the potato growing industry, but with inconclusive outcome (Box 4.3). In Australia's Murray-Darling Basin, groundwater use is conditional on the assessment of third party impacts, an environmental impact assessment, and current and past uses (OECD, 2014b). In France, the authorisation to abstract groundwater is dependent on an impact assessment conducted by the *Préfet* that can be revoked in case of water shortage (OECD, 2010a).

Box 4.3. Well permits and conflicting positions: The case of potatoes in Prince Edward Island, Canada

Prince Edward Island (PEI) has placed a moratorium on high-capacity wells for agriculture irrigation since 2002 while awaiting the results of an impact study on their possible consequence. The moratorium was originally supposed to end in 2003, but has been extended several times. In 2013, studies by the PEI Department of Environment, Labour and Justice showed that the annual recharge rate for groundwater amounted to 2 km^3 per year and that only 0.14 km^3 (7% of recharge) were abstracted. This was accompanied by a change in the calculation of impact on water resource, from a ratio relative to recharge, to a maximum rate of 35% of baseflow (minimum stream coming out of aquifers).

In 2013, following a third relatively dry summer, the PEI Potato Board and Cavendish Farm Inc. used this new information to apply for a lifting of this moratorium in order to use groundwater as supplemental irrigation during dry spells for 30 000 acres of potato fields to a reported maximum volume of 15 million m^3 per year. The Potato Board was supported by the PEI Federation of Agriculture under the objective that it would be needed to ensure that the potato processing industry remain "economically viable" in the future. A study commissioned by the Board estimated that the potato industry accounted for CAD 1 billion of economic activities, representing 9% of the island's gross domestic product.

This application raised opposition from multiple civil-society and environmental groups, including the PEI Wildlife Federation and the PEI Watershed Alliance, supported by biologists and environmentalists from PEI and outside. Opponents argued it would result in groundwater depletion, as well as deter groundwater dependent springs, rivers, and ecosystems. They also worried about increased pesticide run-off, resulting in annual fish kills, and groundwater nitrate contamination and erosion. The Federation of Atlantic Salmon was also cautious and asked for more information.

The Environment Minister brought forth the early study as grounds against the moratorium, but the debate continued after the beginning of the 2014 growing season. The provincial government announced in June 2014 that the moratorium would be maintained until a new Water Act, which would cover the management of all water resources including groundwater, was passed. During the summer of 2014, McCain's Inc. announced it would shut down a major potato processing plant in PEI, and Cavendish Farm threatened to follow suit, noting there was no deadline for this new Water Act. No decision had been made as of September 2014.

Sources: McCarthy (2014); Sharatt (2014); Walker (2014); Wright (2014); Yarr (2014a, 2014b).

Groundwater withdrawal restrictions can take different forms: they can be national, regional or local quotas (sometimes serving as a basis for cap and trade systems); they can apply permanently or in periods of scarcity; vary per year or be fixed; and be either specifically applied to the agricultural sector or to any sector. Countries that implement such restrictions are varied in their general groundwater profile, from Northern Europe to Asia, Mediterranean Europe, as well as North and South America. The specific regulatory designs differ largely.

- In Denmark, groundwater abstraction is limited to 35% of total recharge, but there are additional restrictions including on new irrigation schemes implemented locally by the municipalities (EC, 2012a).
- Farmers in the Nappe de Beauce in France are allocated individual quotas that depend on the local hydrogeological characteristics (Montginoul and Rinaudo, 2013).
- In some provinces of the Netherlands, groundwater withdrawals are allowed only on condition that farms have set up a management plan (OECD, 2010b).
- In the Waikato region of New Zealand, total groundwater extractions are limited to a specified volume, while surface water is limited to a specified percentage (OECD, 2014b). Both limits are fixed by the Waikato Regional Water Plan, a public statutory instrument.
- In Australia's Murray-Darling Basin, the Basin Plan determines environmentally sustainable levels, fixing the overall limit for users each year (OECD, 2014b).
- In several US states, the legislation fixes the overall withdrawal, as observed in the case of the Edwards Aquifer Authority in Texas (Mechlem, 2012), but there are heterogeneous instruments:

- The Upper Republican Natural Resources District in Nebraska, pumping quotas have evolved over time (Fanning, 2012). Allocations are determined for five-year periods with carryover allowed subject to additional constraints, providing flexibility to producers (as water demands vary significantly between wet and dry years).
- Where groundwater is highly connected with surface water, groundwater use is capped to current levels of extractions, with changing surface water quotas. In California, a few districts subject to special acts are allowed to regulate water withdrawals (Hanak et al., 2014).
- In Kansas, Intensive Groundwater Use Control Areas can be introduced specifically to deter stream depletion via restrictions on new permits, withdrawal restrictions, or other relevant measures (Sophocleous, 2010).

Metering can help gauge changes in uses and support a broader assessment for managers. It provides transparency for the public, but also guides users, including farmers in controlling their own use of the resource, potentially in comparison with others. For instance, Portugal has been monitoring groundwater quantity via its piezometric network on a monthly basis and quality on a biannual basis since 1979 (e.g. Ribeiro and Veiga da Cunha, 2010). Water use reporting has been implemented since 1988 in Kansas and the data is used by the local, regional and national agencies to track groundwater use (Sophocleous, 2010). In Victoria, Australia, an Internet-based tool allows for landowners to monitor the state of the resources (Worthington, 2014). Metering is often associated with enforcement measures to remain effective.[5] In Australia, salaried government employees read meters, with large penalties for violators. In some cases, metering is linked to other agricultural policies. Under the Common Agricultural Policy of the European Union, payment of subsidies is contingent of demonstrated compliance with environmental regulations. Producers using groundwater for irrigation must register their wells and install water meters (Montginoul et al., 2014).

Responses from the questionnaire show that mandatory metering is not applied to agriculture wells in Chile and the Czech Republic, and only applied to agriculture wells in the MPW region in the United States, showing that agriculture may be singled out or left out of regulations. Reports are published at least annually in all cases, and monthly in two Japanese and the four Spanish regions. All countries with frequent reports, except Chile, consider that their metering regulation is enforced.

A related issue is that of illegal or unregulated wells that are prevalent in parts of Southern Europe (OECD, 2010a, UNECE, 2011).[6] EASAC (2010) reports that as many as half of the wells could be unregistered or illegal in European Mediterranean countries. Table 4.3 provides a few estimates of the scope of the problem, specifically in agriculture regions.[7] Other countries, like Mexico, have experienced unauthorised use, eased in particular by the use of falsified well and concession registration, and which could represent up to 50% of total concession authorisations in the Valley of Mexico (OECD, 2013b).

Tackling such issues is difficult. In Spain, efforts to complete an inventory and the registration of all wells in the 1990s at an estimated total cost of EUR 66 million led only to partial results in 2001 (Fornes et al., 2007). Some of the river basin management plans, like that of Guadalquivir, explicitly expressed the goal of combatting illegal abstraction, but have not been able thus far to address the problem completely, in part because of the complexities of enforcement and the low impact of fines and legal consequences (ECA, 2014; EEA, 2013). Drought insurance for agriculture is advanced as one possible method to address the issue of illegal wells in European states by discouraging farmers to fight for the last drop (Dionisio and Mario, 2014).[8]

Table 4.3. Estimated number of unauthorised or illegal wells in selected countries

Country	Region	Year	Estimate
Cyprus	National	2012	50 000 boreholes
Italy	National	2006	1.5 million unauthorised wells
	Puglia region	2006	300 000 unauthorised wells
Malta	National	2007	18.5 million m3/year.
Spain	Guadiana Basin	2002	25 000
	National	2005	510 000 wells, 45% of groundwater [1]
	National	2005	90% of farms illegal [3]
	Western La Mancha-Guadiana	2008	22 000 unauthorised boreholes
	Guadalquivir River Basin	2006	10 000

1. Note by Turkey:

The information in this document with reference to "Cyprus" relates to the southern part of the Island. There is no single authority representing both Turkish and Greek Cypriot people on the Island. Turkey recognises the Turkish Republic of Northern Cyprus (TRNC). Until a lasting and equitable solution is found within the context of the United Nations, Turkey shall preserve its position concerning the "Cyprus issue."

2. Note by all the European Union Member States of the OECD and the European Union:

The Republic of Cyprus is recognised by all members of the United Nations with the exception of Turkey. The information in this document relates to the area under the effective control of the Government of the Republic of Cyprus.

3. Different sources.

Source: De Stefano and Lopez-Gunn (2012); Fornés et al. (2007); Hernandes-Mora et al. (2010).

More broadly, there are a growing number of cases where imperfect monitoring of agricultural groundwater use is occurring. In some contexts, groundwater users may not wish to divulge pumping information for strategic reasons (as observed in the west of the United States (e.g. Christian-Smith et al., 2011). This and other reasons have encouraged the use of satellite-based monitoring tools (Castaño et al., 2010, Famiglietti et al, 2011). For example, the Junta Central de Regantes de la Mancha Oriental, a groundwater users' association in the Mancha Oriental region of Spain, charges monthly groundwater use fees to individual producers based on satellite imagery. Each producer is allowed a quota and estimated non-compliance will trigger a site inspection and possible fines (Martin de Santa Olalla et al., 1999; Martin de Santa Olalla et al., 2003). As an alternative, a number of water districts in the High Plains region of the United States have introduced regulations based on historically-irrigated areas, e.g. Nebraska (NE DNR and TPNRD, 2013).

A growing interest in economic approaches, especially market mechanisms

Interest in economic instruments is growing in OECD countries. For example, the National Water Initiative in Australia, an agreement that all states have signed, governs groundwater law. In general, the Initiative requires a move towards economic water management and, as a result, a number of incentive-based groundwater policies have been implemented. These may either be in conjunction with surface water management policies, such as in the Murray-Darling Basin, or independent, such as in the Gnangara Basin (Skurray et al., 2012). Similarly, a number of individual groundwater management areas in the United States, including in California, Kansas, Nebraska, Oklahoma, and Texas, have implemented incentive-based management schemes for agricultural groundwater use (e.g. Wagner and Kreuter, 2004; NE DNR and URNRD, 2010; NE DNR and MRNRD, 2010; Donohew, 2013).

Water pricing is not applied as widely to groundwater as is the case of surface water in OECD countries (OECD, 2010a). Only eight responding countries report applying groundwater charges for pumped groundwater: Denmark, Estonia, France, Israel, Mexico, Portugal, Slovak Republic, and the Czech Republic.[9] The Jeju volcanic island in Korea is the only responding region that applies such charges. Of the eight countries, Mexico and Israel only apply pumping charges to agriculture use, whereas the Slovak Republic does not apply it for abstraction intended for irrigation of agricultural land. France, Mexico and Portugal note that the charges account for the scarcity value of water. Denmark applies a tax on pumped water based on externalities (Calavatra and Garrido, 2010). In the United Kingdom, groundwater charges include an environmental improvement unit charge depending on environmental impacts (EEA, 2013). Six of the 14 districts that regulated groundwater in California pre-2014 legislation have applied charges to pump groundwater, but often with reduced prices for agriculture (Hanak et al., 2014).

The design of pricing schemes, which matters for management, also varies by country (Civita et al., 2010).[10] OECD (2010a) reports that most countries use a fixed fee and volumetric charges for groundwater extraction where the resources are shared with other users. European countries largely use such a scheme with local differences. Belgium includes a volumetric charge above a threshold whose price varies according to aquifer and pollution taxes (OECD, 2010b). The value of charges also differs by location, and use. For instance, groundwater management districts in California charge between USD 18 (Santa Clara Valley) and USD 140 (Orange County) per acre-foot pumped in agriculture and up to over USD 600 to 1000 over a cap (Hanak et al., 2014). Germany applies a charge of EUR 0.0025 – 0.0026 per m^3 used in agriculture in Bremen, Lower Saxony, while the state of Schleswig-Holstein applies a rate 100 times higher, of EUR 0.11/m^3 for non-domestic uses (OECD, 2015c). Similarly, the Czech Republic applies a charge of EUR 0.07/m^3 for drinking water and EUR 0.11/m^3 for other uses (OECD, 2012a).

Several types of *groundwater markets* can be envisioned (chapter 2): trading groundwater entitlements, with or without a cap, but also buying and selling groundwater, or even the land above an aquifer to control its use. Despite the important advantages that groundwater markets can have (Casey and Nelson, 2012; Gardũno and Foster, 2010), there are only a few organised functioning markets. Still, some forms of markets transactions are observed at least on an individual basis in multiple OECD regions.

- Chile, Mexico and Spain allow for entitlements or pumped groundwater to be marketed. Chile is one of the pioneer countries using water markets. Introduced in its Water Act of 1981, surface water markets have had mixed reviews, but less information is available on such mechanisms affecting groundwater. In Mexico, a formal groundwater market has long been in place but has not been very active (Scott and Shah, 2004). Spain introduced groundwater trading in 1999 and has had some useful experience of intra- and inter-provincial markets, although these lacked transparency (Garrido et al., 2012).
- US groundwater regions employ different market approaches.
  - In the NHPA region, entitlements can be marketed and bought out by others, including governments. For instance:
    - The Upper Republican Natural Resources District in Nebraska allows transfers of groundwater pumping entitlements (NE DNR and URNRD, 2010) and there is an adjustment for stream depletion that is unidirectional (i.e. total water use is not allowed to increase even if water moves away from a stream).
    - Several of the Nebraskan Natural Resource Districts of the Platte River Basin allow transfers of the certified right to irrigate acreage (Brozović and Young, 2014).

- Buy-out mechanisms are allowed in the SHPA and MPW regions; conservation-oriented NGOs are increasingly interested in buying groundwater, especially in conservation hotspots (Casey and Nelson, 2012).
  - In Kansas, the Groundwater Management District 5 created an Intensive Groundwater Use Control Area with metering and pumping allocations for which transfers are allowed. A groundwater bank allows such transfers to occur, but they are subject to large conservation offsets and regulatory complexity which represent significant transaction costs.
- In Arizona, farmers are given annual allotments of groundwater use — called "groundwater extinguishment credits" — and they can earn marketable credits for the share of allotments they do not use (Casey and Nelson, 2012; Wichelns, 2010).
- Groundwater markets in western states serve as administrative mechanisms to exchange new groundwater pumping allocations by retiring equivalent quantitative surface water allocations (called "mitigation water") as seen for example in Oregon's Deschutes Water Bank Alliance market (Casey and Nelson, 2012).
- Trading preconditions are also used: some rural areas in California discourage trading of groundwater for out-of-basin export purposes to preserve resources by requiring an environmental impact assessment before the trade is approved (Casey and Nelson, 2012).

• In Australia, groundwater trade is allowed both in terms of permanent water entitlements and temporary water allocations. While all state legislations allow for trading only a few have experienced any trade. Temporary trades have taken place in Queensland and Western Australia, but most have occurred in New South Wales, endowed with large alluvial aquifers, a large number of license, and significant water scarcity constraints (Casey and Nelson, 2012).

• In the Waikato Region of New Zealand, transfers of groundwater permits (entitlements) are allowed under the oversight of the Regional Council. Trading occurs via individual arrangements between entitlement holders. Obtaining a trading allocation requires a new permit or change to the permit and an assessment of the effect of the change, a task that is controlled by the Council (OECD, 2014c).[11]

Even in the absence of actual groundwater markets, transactions can indirectly involve groundwater. Irrigated land buyout is found in Korea and three of the four regions in the United States (NHPA, SHPA, and MPW regions).[12] In Australia's Murray-Darling Basin, third parties are allowed to purchase entitlements in areas where total allocation exceeds what is deemed as a sustainable limit. In the Slovak Republic, land owners must pay charges to use agricultural land for non-agricultural purposes, and these charges are higher for irrigated land. In some countries, cities have bought irrigated lands to ensure groundwater quality (OECD, 2015b). Surface water markets can also be used to replace groundwater markets especially in times of scarcity. During California's 2014 drought, for instance, some farmers bought surface water pumped from aquifers of neighbouring farmers (Sommer, 2014).

Other policies can use economic instruments to indirectly affect the use of groundwater in agriculture, intentionally or not. Agriculture water conservation and irrigation efficiency programs often rely on fiscal instruments to redirect economic incentives towards lower intensive use of groundwater. Figure 4.4 shows that eight of the 21 responding countries report having subsidies for water conservation programs and nine have subsidies for irrigation efficiency. Only a few countries have loans for irrigation efficiency or water conservation purposes. Conditional payments and penalties are less frequently used as an alternative for conservation.

Alternative instruments are also found to encourage irrigation efficiency. In Denmark, a green tax is included on water pumping to encourage increased efficiency (OECD, 2010b). In the four US regions covered in the questionnaire, irrigated land easements are used. Irrigation efficiency can also be linked to US agriculture water conservation programs; in designated areas, groundwater withdrawals are tied to increased irrigation efficiency levels (Schaible and Aillery, 2012).[13]

Figure 4.4. Number of responding countries reporting the use of agricultural water conservation (left panel) or irrigation efficiency programs (right panel) with groundwater effects

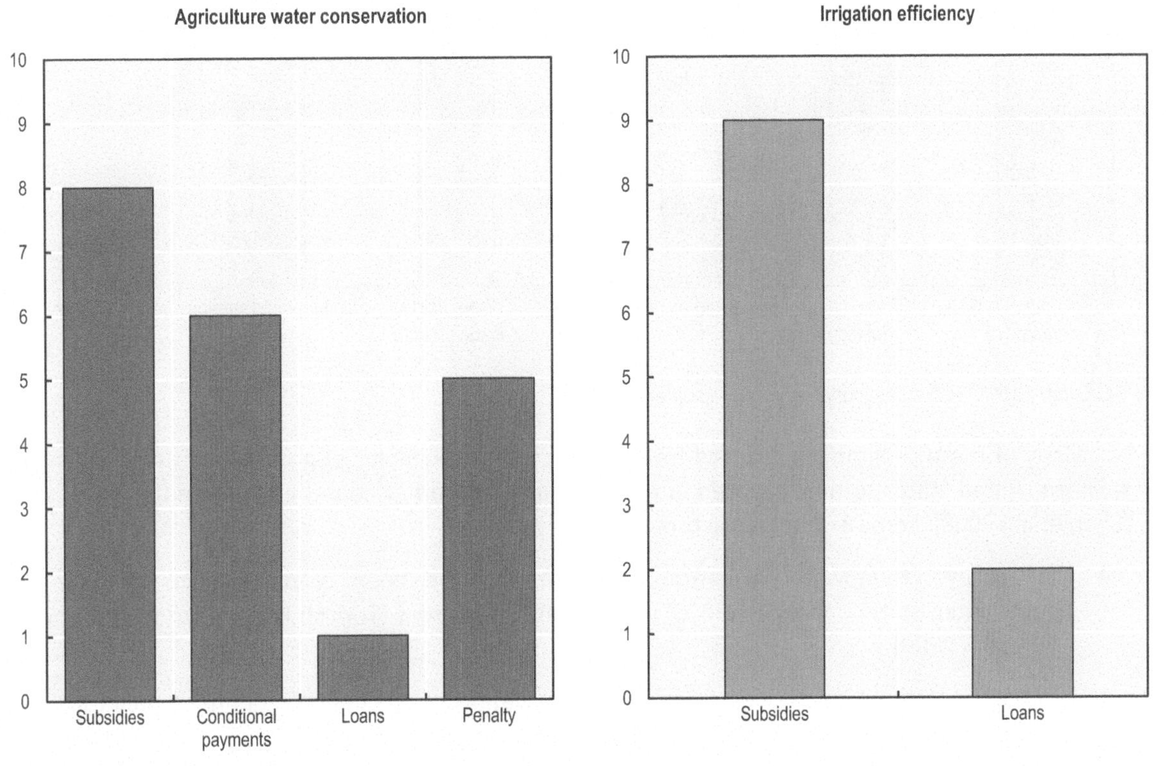

Source: 2014 OECD questionnaire on groundwater use in agriculture.

Partial use of collective management approaches, covering a mixes set of schemes and drivers

The questionnaire proposed three types of drivers and three options to represent the type of collective effort. The first distinction opposes voluntary self-regulations from schemes introduced due to external factors, the regulatory framework or the state authority. The second distinction separates water user associations, district or community-based initiatives, and other institutional mechanisms. Water user associations regroup similar groundwater users (e.g. farmers) already collaborating or contributing to water management districts or community-based mechanisms that are largely based on a geographic and at an administrative level.

As shown in Figure 4.5, only a few countries report the use of such mechanisms, but approaches widely vary. Most schemes are at least framed by regulations and water user associations are in place in most responding countries and regions. Still, responses to the questionnaire are once again diverse; for instance, three significant groundwater irrigating countries use different approaches: Portugal has voluntary groups, Spain has a regulatory framework, and Mexico mandates such groupings. No distinguishable pattern appears; there is no easy correspondence to draw as countries that responded to the first part of the questionnaire did not always do to the second part, and reversely.

Figure 4.5. Number of countries or regions with collective management schemes by drivers (left panel) and scale (right panel)

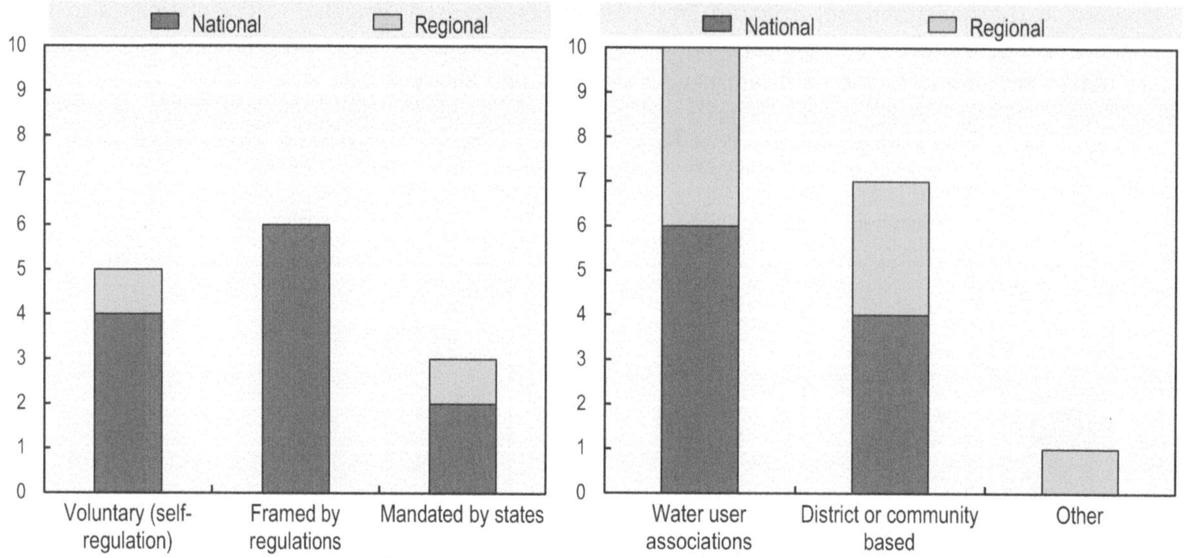

Source: 2014 OECD questionnaire on groundwater use in agriculture.

Many of these programs originated from the successful examples of groundwater districts in the western United States to then be tried in Mexico, Spain and other countries, using different designs (Shah et al., 2008). Some examples are provided below.

- In Turkey, groundwater irrigation co-operatives have been in place since 1966 and have contributed to the management of groundwater resources (Turkish Ministry of Environment and Forestry, 2009).
- In Italy, irrigation boards or associations supply water to users operating under a public law regime and financed by water charges based on cost recovery (Civita et al., 2010).
- In Spain, about 1 400 groundwater user associations are operating either to share the use of wells or groups of wells or to manage groundwater resources associated with an aquifer (Hernandez-Mora et al., 2010). In the United States, irrigation districts, acting as water user associations, are used.
- In California, adjudicated basins, a quasi-collective management-specific setting, allows all users to review groundwater withdrawals and management under the surveillance of a legally appointed Water Master (Cooley et al., 2009).
- In north-west Kansas, a groundwater management district designated "Local Enhanced Management Area" (LEMA) uses a self-regulating scheme with the objective of reducing total water allocation by 20% relative to historic use.

Collective management schemes have been encouraged as alternatives to other management approaches that failed to reach expected results, particularly in developing countries (Iziquierdo et al., 2011; Wester et al., 2009). For instance, technical groundwater committees (COTAS) were introduced in Mexico to protect and restore groundwater bodies under the assumption that top-down programs had failed to address the challenges associated with groundwater depletion, especially in lower income farming communities (OECD, 2013b).

Some local voluntary initiatives have been successful (FAO, 2011). In contrast, the imposition of collective management schemes has led to mixed results. The use of local management districts in parts of the High Plains region of the United States has allowed spatial variations in implemented groundwater use rules to be established that are decided by farmer members of district Boards of Directors and tailored to local needs (e.g. Nebraska DNR and URNRD, 2010; Nebraska DNR and MRNRD, 2010). In Spain, the 1985 Water Act, imposed the creation of water user group to manage the "overexploitation" of aquifers, but in many cases the result has not been successful (Iziquierdo et al., 2011), although some associations have been successful in creating internal mechanisms of abstraction controls (Fuentes, 2011). In Mexico, the COTAS initiative has not been successful in halting groundwater depletion, potentially because of the lack of autonomy it has had in regulating groundwater (Shah, 2008; Wester et al., 2009).

Recent changes in legislation seem to move towards framed collective management approaches. Implementation of the EU Water Framework Directive created the need to form groundwater management bodies to assess and redress the quantitative status of the defined bodies (Iziquierdo et al., 2011). The 2014 groundwater laws introduced in California also attempt to provide an intermediate solution by requesting districts to form local management groups with minimal state role *ex ante*, but potential involvement *ex post* (if no grouping is formed and effective).

Supply-side approaches: An increasing interest in storage options

OECD countries have used several approaches either to supply surface water as a supplement groundwater or to use it as means to recharge aquifers. Figure 4.6 shows responses to the questionnaire, which focus on four main surface-based supply side approaches (OECD, 2011) — expansion of surface water reservoirs, desalination, the use of recycled water, and rainwater harvesting — and three artificial storage options: aquifer storage and recovery (ASR), groundwater banking, and infiltration ponds.

There is increasing interest for storage options in multiple regions. Groundwater banking involves using aquifers as storage capacity for future use; it requires proper hydrogeological conditions and well specified and implemented management objectives to avoid leakages (Maliva, 2013). In some western US states, like Nevada, part of California, and Oregon, groundwater banks serve as a basis of markets, and facilitate transfer and storage (Casey and Nelson, 2012). ASR also involves storing with the purpose of "both augmenting groundwater resources and recovering water in the future for various uses" (US EPA, 2014), but may be more consistent with broader long-term public objectives. The Toscana region of Italy has initiated artificial recharge programs with encouraging results (Civita et al., 2010). Infiltration ponds can be used for banking or ASR as a tool to recharge an aquifer, but they can also be implemented as a means to recharge without a plan monitoring the use and re-use of the resource. Figure 4.6 shows that each of these alternatives is used by at least three countries or regions.

Figure 4.6. Supply-side programs supporting alternative water supplies or storage

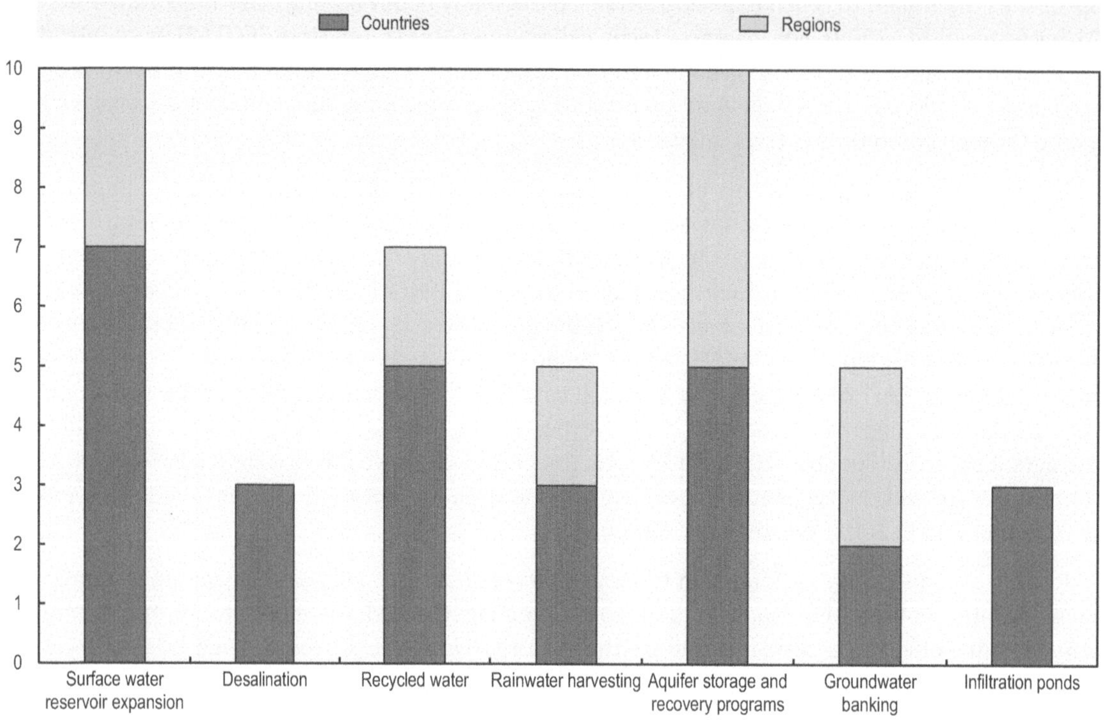

Source: 2014 OECD questionnaire on groundwater use in agriculture.

Supplementing approaches which act more directly on uses vary in scope and outcomes in OECD countries. The development of recycling systems for irrigation is particularly promising in that it does not increase water withdrawals. Rainwater harvesting can also provide simple solutions. Israel is using treated wastewater to recharge groundwater and for irrigation (OECD, 2012c), and has developed water harvesting in individual households to combat water scarcity (Ronen et al., 2012). On the other hand, the Spanish experience with desalinated water suggests that prices and final quality matter greatly in the outcome for aquifers. In the Campo de Nijar region, the provision of desalinated water has not effectively reduced groundwater pumping for irrigation, even under the constraint of increasing salt water intrusion (Lopez-Gunn et al., 2012a). In the Canary Islands, the combination of publically-provided desalinated water and private groundwater ownership has helped manage groundwater pressures (Custodio and del Carmen Cabrera, 2012). The Arvin Edison Water and Storage District, set up in 1942, in Kern county in California's Central Valley, has engaged in conjunctive use involving groundwater banking during wet years and pumping back during dry seasons or years at an annual benefit of USD 488,000, or 47% of the value of groundwater (Schoengold and Zilberman, 2007; Wichelns, 2010)

Some of these instruments can be combined, as seen in the case of multiple urban-rural partnerships, aiming at reducing pressure on groundwater or payments for farmers for supplying recharge (Box 4.4) (see also OECD, 2015b).

Alternative approaches are found in specific countries, sometimes mixing different instruments. In Belgium, one of the policy objectives is to discourage drainage to encourage groundwater recharge in upstream area and prevent downstream flooding (OECD, 2010b). In California, research is conducted to see how directing excess water to dormant fields in winter could be used to maximise groundwater recharge (Harter and Dahlke, 2014). In Almeria and Alicante, Spain, modernisation plans have been implemented with the dual purpose of increased water efficiency and limited leakages and the use of alternative sources to reduce the intensity of groundwater use (Lopez-Gunn et al., 2012b).

Box 4.4. Rural-urban water and financial transfers to address groundwater overdraft

A number of local management initiatives have been developed around cities with the objective of finding ways to conserve groundwater supplies.

First, in multiple cases, the use of treated wastewater has been offered by cities to irrigators in exchange for groundwater, with the purpose of conservation (or banking). This has been operated, for instance, in Topeka, Kansas, where the city paid for groundwater rights in exchange of treated municipal wastewater. The conservation objective was fulfilled and the city reportedly did not use groundwater for seven years. The city of Wichita Kansas, facing the risk of groundwater depletion embarked on a large transfer of surface water into its section of the High Plains Aquifer. Similarly, Dodge City, Kansas, provides treated wastewater for irrigation in exchange for groundwater rights. In Santa Clara, California, a conservation district was created and managed to stop land subsidence that damaged infrastructures. The district's management plan included monitoring of groundwater use among rural and urban users, importing surface water, and artificially replenishing the aquifer with treated wastewater. Plans have also been developed to reduce groundwater pumping-induced intrusion of saline water into the aquifer via collective management around Bordeaux in France, via the provision of treated wastewater mixed with surface water for irrigation in the Salinas valley of California, or in Toscana, Italy by using treated urban and industrial wastewater. Tucson, Arizona has developed an ASR program with the two-fold purpose of storing water underground for future use while replenishing the already pumped groundwater.

Other cases have involved payments or financial transfers. Farmers have been paid by cities and industrial companies to encourage practices that help recover the aquifer, such as paddy rice flooded practices. The city of Ono in Fukui Prefecture, which received the Japan Water Grand Prize in 2012, was one of the early adopters of groundwater recharge via paddy fields storage in the late 1970s. A similar mechanism was introduced in Kumamoto, Japan, where the city acted in conjunction with Sony Corporation and a local foundation to support farmers (Hashimoto, 2013). The rice was then promoted as environmentally friendly. The city of Azumino set up a mechanism of pay-for-use to ensure conservation for the benefits of urban, industrial and agricultural users. Other cities have used conservation easements, providing protection to land surrounding the city in exchange for aquifer recharge (e.g. San Antonio, Texas).

Source: Peck (2007) ; Borchers et al. (2014) ; Barraqué et al. (2010) ; Groot, (2013) ; Civita et al. (2010); Hashimoto (2013); Lee (2014).

Other sectorial policies affecting groundwater use

Few countries tax or subsidise electricity

Groundwater irrigation in OECD countries relies almost entirely on electric or, in fewer cases, on diesel pumps. This implies that groundwater development is dependent on the availability of affordable connection to electric sources.[14] As explained in Chapter 2, subsidies in India and Mexico have contributed to support the overdraft of groundwater resources. So far, several barriers have prevented reforming these multi-decades programs, but there have been efforts to move in this direction. An innovative scheme in the Indian State of Gujarat, based on decoupling electricity networks, succeeded in reducing the use of groundwater (Shah et al., 2008) and a few pilot cases have been introduced in Mexico with the intention to evaluate how such a reduction could be implemented (De Richter, 2013).[15]

The questionnaire included three options of programs for electricity – subsidies, taxes, other energy programs – with limited responses. Chile, Portugal and the Korean region of Jeju volcanic island use electricity subsidies and other programs and the Netherlands employ the three options at the national level. Portugal's program focuses specifically on increasing energy efficiency in agriculture irrigation. France, Mexico, and Spain also report employing electricity taxes at the national level.

Half of respondents apply land policies with implications on groundwater use

The close linkages between land ownership and groundwater use (and ownership) make land policies potentially important when considering groundwater management. The questionnaire included three main options: zoning, regional allocation systems, urban-rural collaboration (Figure 4.7).

Over half of the 21 responding countries report using zoning restrictions on groundwater use, in some cases potentially for water quality preservation (drinking water), but clearly affecting access to aquifer and therefore groundwater use (Figure 4.7). Indeed, zoning mechanisms are considered major features of recent groundwater legal regimes, ranging from re-charge protection zones to zoning for drinking water (Mechlem, 2012). These are found at the national and regional levels in France and the Netherlands (six regions in Figure 4.7). In France, under the 2006 *Loi sur l'Eau et les Milieux Aquatiques*, the *Prefets* can create zones of environmental constraints on which they can impose water restrictions (Barraqué et al., 2010).

Figure 4.7. Number of responding countries and regions with land policies related to groundwater

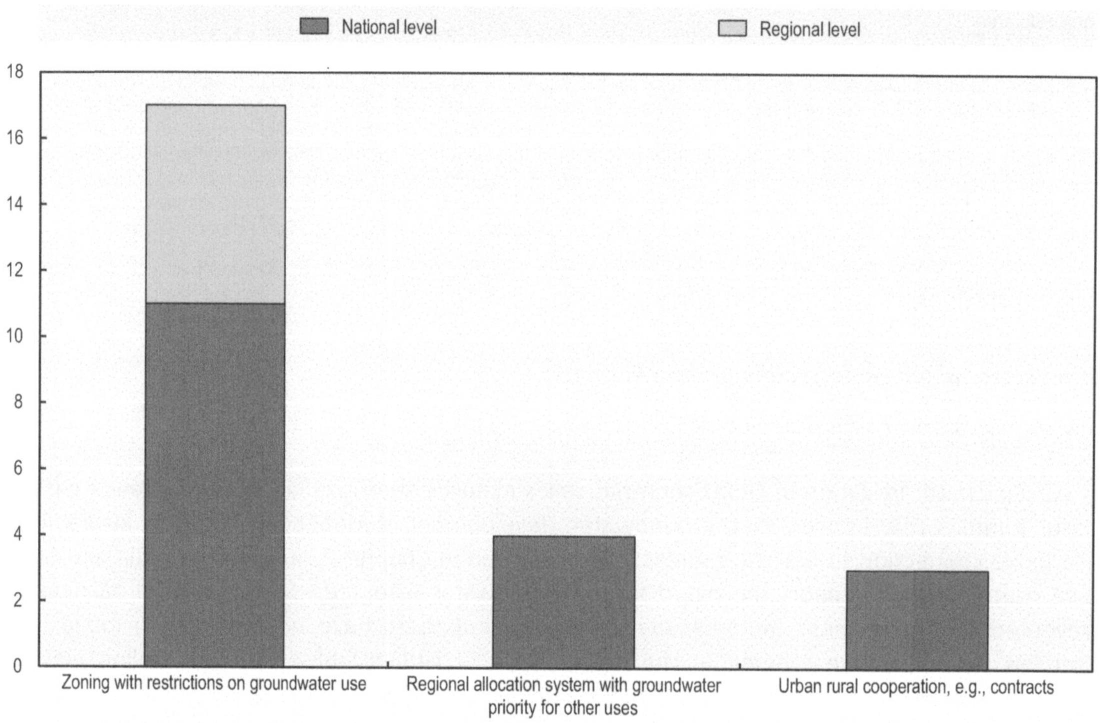

Source: 2014 OECD questionnaire on groundwater use in agriculture.

Land retirement programs can also be used in schemes intended to redress externalities as observed in Nebraska, United States. In recent years, the Upper Republican Natural Resources District has had an active land retirement program, spending USD 10 million to purchase 3 300 acres (1 300 hectares) of irrigated land in 2011 (in the Rock Creek area) and joining with the Middle and Lower Republican Natural Resources Districts and the Twin Platte Natural Resources District to purchase almost 20 000 acres (8 100 hectares) of land for USD 83 million in 2012 (McCabe, 2013). In both cases, the Natural Resources District is constructing stream augmentation projects that will link wells to nearby streams directly with a pipeline. This will allow pumping of groundwater directly into the river to provide compliance with interstate surface water compacts in drought years.[16]

Four OECD countries report having regional allocation systems and Estonia, Israel and Portugal report implementing urban-rural co-operation schemes on land affecting groundwater uses, but other land-related policies have been implemented. In Slovenia, a subsidy program funded by the European Rural Development Fund supports moving towards land use that is less water-demanding (EC, 2012b). In the United States, the government funds and administers a number of agricultural irrigated land-fallowing programs as incentives to reduce water consumption (Casey and Nelson, 2012).

Several watershed conservation programs indirectly affect groundwater use

Watershed conservation programs may also have indirect effects on groundwater use. These encompass regulatory and economic instruments. Exclusion zones for conservation purposes and limits on groundwater use near protected areas are prevalent in nine and seven of the 20 responding countries, respectively. Spatial approaches can in particular help address local externalities. Five countries (Estonia, France, Korea, Slovak Republic and Mexico) and four regions (Almonte-Marismas in Spain, and NHPA, SHPA and MPW in the United States) use groundwater entitlement acquisition for conservation purposes.

Programs pertaining to wetlands in particular are bound to be drivers of groundwater conservation. In Japan, Korea and the United States, for instance, there are programs whose objectives is the conservation of wetlands while reinforcing the groundwater recharge capacity of agriculture, including paddy rice system (OECD, 2010b).

A growing set of climate change adaptation plans, some drought insurance programs

As emphasised in Chapter 1, groundwater can serve as a mainstream instrument for adaptation. Agricultural adaptation policies and plans are diverse and consider supply, demand, and resilience to extreme events (OECD, 2015a). Some of these policies explicitly rely on water (OECD, 2014a). If some policies explicitly consider groundwater policies specifically, others look at other aspect of the agriculture-water-climate change nexus.

Research and development leading to practices and technologies that are better adapted to climate change may result in reduced groundwater use. Sophocleous (2012) notes that a USDA report comparing multiple options to manage the High Plains Aquifer found the best outcomes with the use of yield-increasing biotechnology (assumed to be 5% per decade) coupled with groundwater restrictions (10% less water per decade). Groundwater modelling and data can also contribute to better management of groundwater, thus supporting adaptation. Figure 4.8 (upper panel) shows the frequency of four types of adaptation investment programs among responding countries.

The lower panel of Figure 4.8 shows that while seven countries have private insurance mechanisms, only Chile, the United States (plans for field crops), Spain and France support government-based insurance programs.

A number of drought policies also refer explicitly to groundwater. In most OECD countries, a prioritisation system is implemented to allocate surface and groundwater resources in times of drought (OECD, 2015c). In Sweden, provincial governments can limit irrigation during droughts. In the US state of Georgia, the use of minimum stream flow can serve as a trigger for groundwater irrigation restrictions (Wright et al., 2012).

Figure 4.8. Number of responding countries with climate change adaptation (upper panel) and drought insurance programs (lower panel)

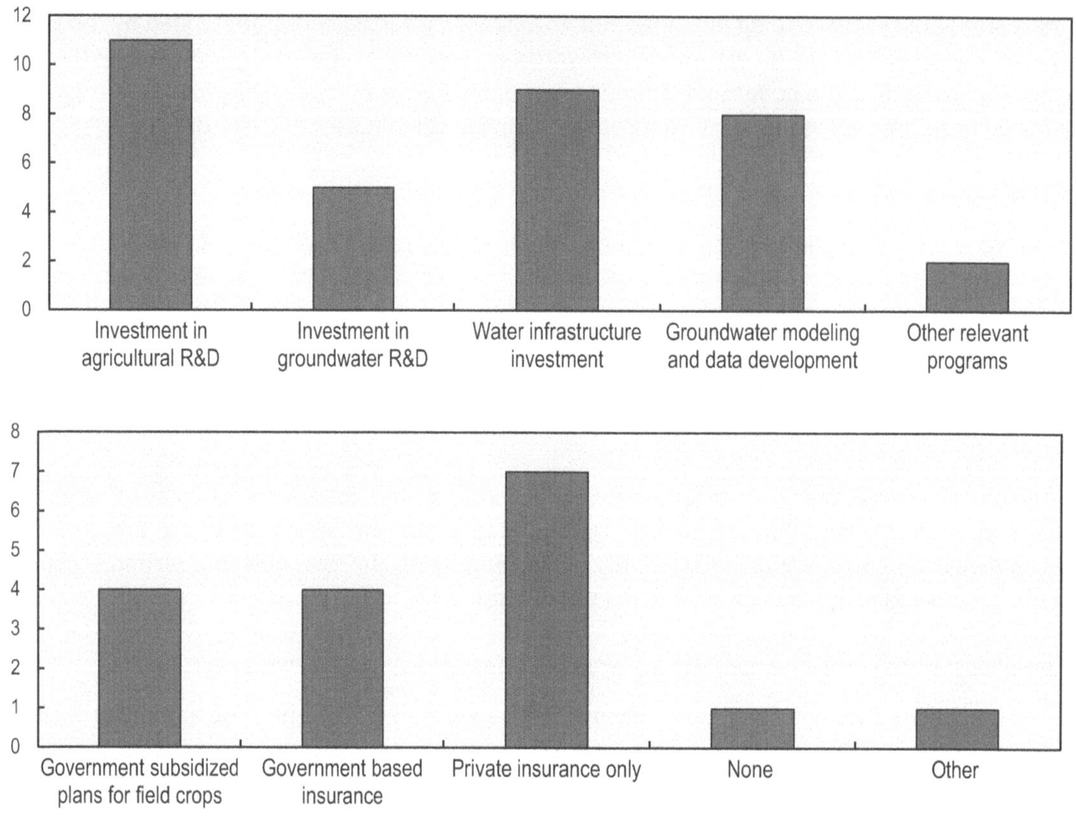

Source: 2014 OECD questionnaire on groundwater use in agriculture.

Farm support policies generally not biased towards water intensive crops

Responses from the questionnaire report that OECD countries that support their farmers do not generally use this type of incentive, but there are a few exceptions. Still, Portugal notes that under Pillar 1 direct payments of the Common Agricultural Policy (CAP), some water intensive crops are supported (among others), and that some remaining coupled payments do apply to livestock systems and some intensive systems. Four countries implement biofuel production support programs (the Czech Republic, the Netherlands, the United Kingdom and the United States), which could in certain cases be detrimental to groundwater resources in the long run (Chapter 3).

Some subsidised programs are designed to support more sustainable management of groundwater. Austria, France, and the Slovak Republic include that agro-environmental measures may, on the other hand, reward farmers for better practices *vis-à-vis* groundwater. In Portugal, some of the rural development programs under Pillar 2 of the CAP encourage better practices or investment that contributes to the recharge of aquifers; e.g., via agroforestry, or the installation of riparian strips.

Are policy instruments corresponding to specific groundwater characteristics? Findings from a regional analysis

The previous section has demonstrated the diversity of management and policy approaches used across OECD countries and regions, but an underlying question is whether this diversity reflects the differences in groundwater characteristics. The data obtained from the questionnaire on groundwater regions, which requested specific information on the characteristics of groundwater resources and use (Chapter 2) on the one hand, and management and policy approaches (chapter 3) on the other, provides an opportunity to explore the potential presence of such linkages. As shown in Table 4.4, the data were obtained on both sides for 20 regions in 11 OECD countries. The corresponding countries also provided detailed responses on national policies that apply at the regional level.

Available responses from these 20 regions were used to form aggregated indices representing bundles of indicators in each area of interest for qualitative cross-region comparisons. Annex 4.A provides the overall computation procedure, the goal of which was to obtain a set of indicators that for the first part (characteristics following chapter 2, Table 2.3) consistently increase with the likelihood of groundwater stress (and conversely), and for the second part (policies) increase with the scope of policy responses to control groundwater use (and conversely decreases with policies that do not encourage control or even support groundwater use).

These indicators were compared via pairwise comparison and then used for region-to-region comparison. The goal of this exercise is not to derive meaningful absolute values of stress and policy coverage, but rather to enable a comparison of the relative importance of the types of constraints and approaches used among regions.

Annex 4.B provides detailed results of the analysis (indicators and region-by-region comparison). First, the characteristics of groundwater bodies show a great diversity. Four regions stand out in terms of high climatic constraints, groundwater use, relatively lower availability of surface water, competition with other uses and externalities (but with differences in aquifer characteristics): the Mexican Region Laguna, Israel's Western Galilee, the US MPW region, and the Australian Murray-Darling Basin. A second group of regions in Japan, Korea, Denmark, France, Portugal, Spain and the Netherlands do not rank as high in many of these categories with relatively more available surface water, less arid current and projected climate, lower level of groundwater use and externalities (with the exception of Spain's Mancha Occidental). Lastly, the Italian Campania region and the NHPA, SHPA and MAA regions in the United States appear to be intermediate in these dimensions. Regions differ in aquifer characteristics, but there is no simple pattern.

These bundles of regions do not fully correspond with those observed when considering policy responses. Israel's Western Galilee and Mexico's Region Laguna both stand out again in terms of their relative emphasis on supply-driven approaches. The severity of freshwater pressures may have pushed these countries towards costly but seemingly unavoidable supply investments. On the demand side, while Israel's Western Galilee uses regulatory approaches, Mexico's Region Laguna stands out in its significant use of economic and collective management approaches. Trends in the other regions are not as clear cut and tend to reflect national specificities.

- In the United States, the MPW, SHPA, and MAA regions share a number of characteristics in that they each provide a legislative framework with relatively higher freedom to operate and limit groundwater regulations, but policies that may indirectly regulate groundwater use. The NHPA region has a higher use of economic, regulatory, and collective management approaches.
- Australia's Murray-Darling Basin is distinguished by the relatively high level of economic instruments, moderate use of supply side approaches, and lower use of regulatory instruments.
- The three Japanese regions apply the same types of approaches, based mostly on national legislation, with a similarly relatively high level of freedom to operate, relatively lower regulations, and very limited measures supporting groundwater use.

Table 4.4. Regional coverage of the OECD questionnaire

Country	Region	Notation	Localisation	Characterisation of groundwater resources	Policies
Australia	Great Artesian Basin	AusGAB	North	X	
	Murray-Darling Basin	AusMDB	Central-South	X	X
Denmark	Western Jutland[1]	DenWJ		X	X
France	Nappe de la Beauce	FraNB	Ile de France	X	X
	Departement de la Vienne	FraDV	Poitou-Charentes	X	X
Israel	Western Galilee	IsrWG		X	X
Italy	Puglia	ItaP	Conzorzio di Bonifica Arneo	X	
	Campania (Ufita)	ItaC	Conzorzio di Bonifica UFITA	X	X[2]
Japan	Kinugawa Seibu	JapKS	Kanto district	X	X
	Noubiheiya Seibu	JapNS	Tokai district	X	X
	Kikuchi Heiya	JapKH	Kyushu district	X	X
Korea	Jeju volcanic island	KorJvl		X	X
Mexico	Region Lagunera	MexRL	Cohuila and Durango states	X	X
Netherlands	North-Brabant	NldNB	Meuse River Basin, South of the Netherlands	X	X
	Limburg	NldL		X	X[2]
	Gelderlan and Overjissel[3]	NldG	Rhine-East River Basin, East of the Netherlands	X	X[2]
Portugal	Tejo e Ribeiras do Oeste River Basin District	PorTRO	Ribatejo e Oeste, Beira Interior and Alentejo regions	X	X[2]
Spain	Mancha Occidental I	SpaMOc	Ciudad Real, Castilla La Mancha	X	X
	Campos de Montiel	SpaCM	Ciudad Real, Albacete, Castilla La Mancha		X
	Almonte -Marismas	SpaAM	Huelva y Sevilla, Andalucía		X
	Mancha Oriental	SpaMOr	Valencia, Albacete, Cuenca. Castilla La Mancha y Comunidad Valenciana		X
Turkey	Küçükmenderes Basin	TurKB		X	
United States	Northern High Plains Aquifer	NHPA	North Dakota, South Dakota and Nebraska	X	X
	Southern High Plains Aquifer	SHPA	Kansas, Oklahoma, Texas	X	X
	Mississippi Alluvial Aquifer Region	MAA	Arkansas, Mississippi, Louisiana and a corner of Missouri	X	X
	Mountain and Pacific West Region	MPW	Washington, Oregon, California, Idaho, Utah, Nevada, Montana, Wyoming, Colorado and New Mexico	X	X
TOTAL		27		23	23

1. The assessment is not done at the municipal level.
2. National policies are used instead of region-specific responses (others combine region specific and national policies).
3. The two regions have identical responses and will therefore be represented as one.

Table 4.5. Selected indicators of groundwater resource characteristics and policies by regions

Characterisation of groundwater region		Management and policy approaches	
Name	Description	Name	Description
IndClim	Agro-climatic conditions (current and projected)	**IndEntFTO**[1]	Freedom to operate based on ownership and entitlement characteristics
IndAq	Aquifer type and geology	**IndRegGW**[1]	Regulations of groundwater (management plans, wells, monitoring, withdrawals)
IndSW	Surface water availability and use	**IndEconGW**[1]	Economic instruments to control groundwater use (charges, markets, buyouts)
IndGWuse	Groundwater use in agriculture (current and projected)	**IndCMGW**[1]	Collective management approaches to control groundwater
IndOtheruse	Competition from other users (current and projected)	**IndSupply**[1]	Supply side approaches
IndExt	Externalities due to groundwater pumping (Current and projected)	**IndOtherControl**[1]	Other indirect controls (energy, agriculture and conservation programs)
		IndOtherConso[1]	Other programs that may indirectly support groundwater use (energy, agricultural)

1. The indicators account for policies and programs set at the national level.

- The French Department of Vienne and Nappe de la Beauce regions both rely on a relatively high level of collective management, economic instruments and other controls, limited freedom to operate, and regulatory measures.
- The Netherlands' regions report moderate regulations and, like Denmark's Western Jutland, provide a relatively low level of freedom to operate.
- Spain's Mancha Occidental and Korea's Jeju volcanic Islands lead the pool in regulations, together with a moderately high level of common management approaches and economic instruments. Portugal's Tejo e Ribeiras do Oeste also shares with Spain's Mancha Occidental a high level of other control policies.

Apart from national or continental similarities and the relationship between higher stress and supply-side approaches, there is no simple correspondence between the characteristics of groundwater resources and use and policies in these agriculture groundwater regions. Still, pairwise correlations (Annex 4.A) indicate that:

- The freedom to operate indicator is positively correlated with groundwater use in agriculture and competing demand related stresses, and negatively correlated with surface water use; it is larger in high groundwater using regions, where farmers can decide when to pump.
- Economic approaches are more prevalent in regions with high climatic stress, low surface water, and more significant externalities.
- Common management approaches are more prevalent in areas facing externalities;
- Supply-side approaches are negatively correlated with the availability of surface water, and positively correlated with climatic stress, externalities and weakly with the aquifer index.
- Other indirect control approaches are weakly correlated with the indicator for externalities.
- Policies supporting groundwater irrigation are positively correlated with groundwater use.

Naturally, these pairwise correlations are only relevant for this subset of regions, may not stand up to more robust statistical analyses, and do not, in any case, support causality in one way or another. Still, they support the region-to-region analysis (Annex 4.B) and suggest potentially interesting links to empirically validate such links as between entitlement characteristics (freedom to operate) or support policies and groundwater use.[17] Analysing whether local groundwater characteristics drive these policies or whether policies impacted these characteristics would be the other important question.

More generally, these results suggest that the choice of certain management options may indeed be at least partially related to characteristics of groundwater use. In particular, economic and supply side approaches might be more prevalent in agriculture areas with higher groundwater stress.

At an higher aggregate level, one can use these indicators to derive two mega-indicators capturing the overall relative groundwater stress and the relative number and diversity of instruments to manage agricultural groundwater use by region.[18] The result can be seen in Figure 4.9, with the stress indicator represented by the horizontal axis and the policy indicator on the vertical axis.[19]

Several important caveats should be emphasised before interpreting this figure. First, none of the numbers have any meaningful value. Second, the figure separates the stand-out approaches rather than those with more stress or approaches overall. Third, regions with an incomplete response may be at least partially misplaced. Finally, the policy axis stands more for the relative number of approaches towards controlling groundwater than for their stringency or effectiveness.

Acknowledging these limitations, Figure 4.9 does provide a set of possible bundles of regions with similar constraints, characteristics, and/or management approaches, consistent with the above pooled comparison of regions. If there are three relative clusters of groundwater stress on the horizontal axis (shown in the segments at the top: seven regions are well on the left of the axis, five on the right, and the remaining eight in the middle), they translate into five policy/characteristics clusters, represented by circles. In particular, the Western Galilee and Laguna Region stand out positively in both dimensions, and the Murray-Darling, MPW and SHPA regions score both relatively high on challenges but with less hands-on policy approaches. The MAA, Portuguese and Japanese regions have relatively low scores in the two components, and the Dutch, Danish and Nappe de Beauce regions have lower stress and a higher index of management. Korean, Spanish and NHPA regions are presented with relatively moderate to high level of management and constraints.

Interpreting the vertical axis as a proxy for intervention, Figure 4.9 suggests that countries and regions have used different degrees of interventions to respond to locally specific stresses. Interestingly, some regions employ relatively high degrees of oversight with much lower constraints, and other regions have a lower oversight with higher constraints. Intervention does not guarantee success and, as argued earlier, some may be perfectly unjustified economically. But there is no evidence of an evolution in public policy interventions with the severity of stresses.

Figure 4.9. Comparing relative groundwater stress and policy approaches in the responding regions

Horizontal axis: Relative agro-climatic and groundwater stresses

Vertical axis: Relative importance of groundwater use management approaches

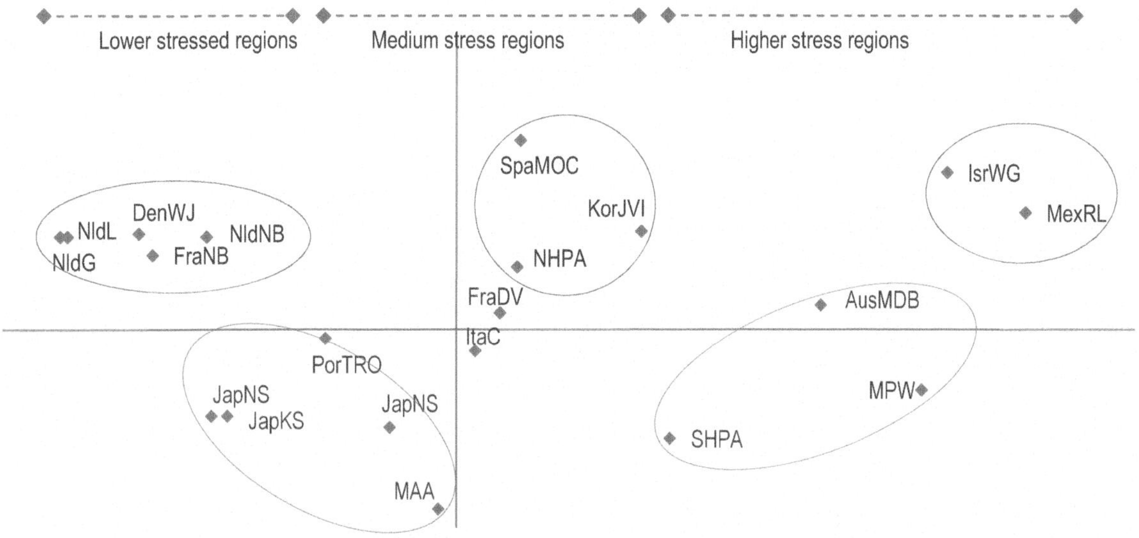

Note: AusMDB: Murray-Darling Basin; DenWJ: Western Jutland, FraNB: Nappe de la Beauce, Fra DV: Departement de la Vienne; IsrWG: Western Galilee; ItaC: Campania (Ufita); JapKH: Kikuchi Heiya; JapKS: Kinaguwa Seigu; Jap NS: Noubiheiya Seigu; KorJvl; Jeju volcanic Island; MexRL: Region Laguna; NldG: Gelderlan and Overjissel ; NldL: Limburg NldNB; North Brabant; PorTRO: Tejo e Ribeiras do Oeste; SpaMOC: Mancha Occidental; MAA: Mississippi Alluvial Aquifer; MPW: Mountain and Pacific West; NHPA: Northern High Plains Aquifer; SHPA: Southern High Plains Aquifer.

Source: Derived from results from the 2014 OECD questionnaire on groundwater use in agriculture (see Annexes 4.A1 and 4.A2).

Notes

1. The full 2014 OECD questionnaire on groundwater use in agriculture can be obtained by contacting tad.contact@oecd.org.
2. Regional responses are described only for those that differ from the national response and for the four regions of the United States (for which entitlement systems are set at state level).
3. Several countries have attempted to change ownership from private to public. Spain's experience following its 1983 Water Act has shown that implementing this type of reform is challenging, especially in areas with past intensive use of groundwater for irrigation by thousands of farmers (Llamas and Garrido, 2007).
4. Groundwater management is also done at the regional level in Belgium (OECD, 2010b).
5. An example of an effective metering and enforcement scheme is the Upper Republican Natural Resources District in Nebraska. In 2010, this district revoked groundwater-pumping entitlements with an estimated value in excess of USD 3 million for several groundwater users who had attempted to increase their water use illegally by bypassing their well flow meters (McCook Gazette 2010).

6. Another growing issue is that of abandoned wells which can affect flows and create quality problems in the long term (personal conversation with T. Jarvis, Oregon State University).

7. In addition to the countries reported in Table 4.3, Greece and Mexico are known to have a significant number of illegal wells (de Stefano and Lopez-Gunn, 2012; de Richter, 2013).

8. However that presumes the insurer is willingly accepting to take on the moral hazard risk that the farmers represent.

9. According to Shah (2008), Israel employs the "best example of water pricing for agriculture use". Other sources report that Germany charges all users except for those in agriculture (OECD, 2015), and Turkey uses a flat rate per hectare (OECD, 2012b).

10. Flat pricing over time does not affect long term incentives (Civita et al., 2010). This type of scheme may also affect the response to pricing; e.g. a two-layer block-rate tariff with a very low price up to a certain quantity, and very high above it may have similar effects to that of a water quota, as observed in Mexico (de Richter, 2013).

11. Estonia also reports allowing pumped water to be marketed, although with limited implications for agricultural purposes.

12. Systems of this nature operate in the Twin Platte and Central Platte Natural Resources Districts in the Platte River Basin in Nebraska (Young, 2014).

13. Under the 2014 US Farm Bill, a Regional Conservation Partnership Program has been introduced to encourage private-public initiatives for conservation purposes, including agricultural groundwater conservation.

14. Intensive use of groundwater can also exert pressure on electric power generations. In the United States, it is estimated that groundwater pumping for agriculture represents about 1% of total electricity use (Water in the West, 2013). In Mexico, it represented about 6% of electricity demand in 2001 (Scott and Shah, 2004).

15. In 2008, the Spanish government removed electricity subsidies for irrigation as part of the liberalisation of the electricity market, leading to an 60% increase in cost, which resulted in some irrigation being abandoned (Calatrava and Garrido, 2010; Fuentes, 2011).

16. Note that as pumping itself induces stream depletion, stream augmentation is a temporary measure that has been controversial within the community as it can be viewed as depleting the limited groundwater resource without increasing agricultural production through irrigation.

17. Lopez-Gunn and Llamas (2008) report, for instance, that when looking at past experience internationally, it is difficult to attribute any clear advantages from management between privately- *versus* publically- owned groundwater.

18. This was done by compiling weighted averages of the derived indicators. To do so, a -1 weight is applied for surface water availability (IndSW) and +1 for all other indicators of groundwater system characteristics, and similarly a -1 is used for groundwater use indirect supporting measure (IndOtherConso) and freedom to operate (IndEntFTO) and +1 for all other variables on the management side.

19. The combined index of stress is specific to agriculture and groundwater, and accounts for future changes in climatic conditions. More general indexes of freshwater stresses are available as part of the OECD environmental indicators (OECD, 2013a).

References

Barraqué, B. et al. (2010), "Groundwater in the Southern Member States of the European Union : an assessment of current knowledge and future prospects. Country report for France", European Academies Science Advisory Council (EASAC) Halle.

Borchers, J.W. et al. (2014), "Land Subsidence from Groundwater Use in California", Ludhorff and Scalmanini Consulting Engineers, Sacramento, CA.

Brozović, N. et al. (2006), "Optimal management of groundwater over space and time", in R. U. Goetz and D. Berga (eds.), *Frontiers in water resource economics*, Springer, New York.

Brozović, N. and Young, R. (2014), "Design and Implementation of Markets for Groundwater Pumping Rights", in K. Easter and Q. Huang (eds.), *Water Markets for the 21st Century: What Have We Learned?*, Springer, New York.

Calatrava, J. and A. Garrido (2010), "Measuring irrigation subsidies in Spain: An application of the GSI method for quantifying subsidies", Global Subsidies Initiative, International Institute for Sustainable Development (IISD), Geneva, Switzerland. Available at: http://www.iisd.org/gsi/sites/default/files/irrig_spain.pdf.

Casey, M. and R. Nelson (eds.) (2012) "Groundwater Science, Policy, Partnerships, and Markets", In *Groundwater Science, Policy, Partnerships, and Markets*, Stanford University, Palo Alto, CA.

Castaño, S. et al. (2010), "Methodology for Quantifying Groundwater Abstractions for Agriculture via Remote Sensing and GIS", *Water Resources Management*, Vol. 24 (4), pp. 795-814.

Christian-Smith, J. et al. (2011), "Impacts of the California Drought from 2007 to 2009", Report of the Pacific Institute, Oakland, CA, http://www.circleofblue.org/waternews/wp-content/uploads/2013/04/ca_drought_impacts_full_report.pdf

Civita, M.V. et al. (2010), "Groundwater in the Southern Member States of the European Union: an assessment of current knowledge and future prospects: Country report for Italy", European Academies Science Advisory Council (EASAC), Halle.

Cooley, H et al. (2009), "Sustaining California Agriculture in an Uncertain Future", the Pacific Institute, Oakland, CA.

Custodio, E. (2010), "Intensive groundwater development: a water cycle transformation, a social revolution, a management challenge", in L. Martinez-Cortina, A. Garrido, and E. Lopez-Gunn (eds.), *Water Workshop on Re – thinking Water and Food Security Paradigms*, CRC Press, London.

Custodio, E. and M. del Carmen Cabrera (2012), "The Canary Islands", in L. de Stefanoand, M.R. Llamas (eds.), *Water, Agriculture and the Environment in Spain: Can we square the circle?*, CRC Press, London.

Delaware River Basin Commission (2008), *State of the Delaware River Basin Report*, West Trenton, New Jersey.

De Richter, C.V., (2013), "Green growth challenges and the need for an energy reform in Mexico", Economics Department Working Paper No 1095, OECD, Paris, France.

De Stefano, L., and M.R. Llamas (2012), *Water, Agriculture and the Environment in Spain: Can we square the circle?*, CRC Press, London.

De Stefano L, and E. López-Gunn (2012), "Unauthorized groundwater use: institutional, social and ethical considerations", *Water Policy*, Vol. 14, pp. 147–160.

De Stefano, L. et al. (2013), "Easier Said Than Done? The Establishment of Baseline Groundwater Conditions for the Implementation of the Water Framework Directive in Spain", *Water Resources Management*, Vol. 27(7), pp. 2691–2707.

Dionisio, P. and G. Mario (2014), "Insuring water: a practical risk management option in water-scarce and drought-prone regions?" *Water Policy*, Vol. 16 (2), pp. 244-263. doi:10.2166/wp.2013.131

EASAC (European Academies Science Advisory Council) (2010), "Groundwater in the Southern Member States of the European Union", EASAC Policy Report 12, EASAC, Halle.

EC (European Council) (2012a), "Commission Staff Working Document – Member State: Denmark, Accompanying the Report from the Commission to the European Parliament and the Council on the Implementation of the Water Framework Directive (2000/60/EC) - River Basin Management Plans", European Commission, Brussels.

EC (European Council) (2012b), "Commission Staff Working Document - Member State: Slovenia, Accompanying the Report from the Commission to the European Parliament and the Council on the Implementation of the Water Framework Directive (2000/60/EC) - River Basin Management Plans", European Commission, Brussels,.

EC (European Council)(2006), "Directive 2006//EC of the European Parliament and of the Council on the protection of groundwater against pollution and deterioration", European Parliament and the Council, Brussels.

EC (European Council)(2000), "Directive 2000/60/EC of the European Parliament and of the Council establishing a framework for the Community action in the field of water policy", European Parliament and the Council, Brussels.

ECA (European Court of Auditors) (2014), "Integration of EU water policy objectives with the CAP: a partial success", Special Report, ECA, European Union, Brussels.

EEA (European Environmental Agency) (2013), "Assessment of cost recovery through water pricing" EEA, Copenhagen.

EUWIMed (European Union Water Initiative-Mediterranean region)(2007), "Mediterranean Groundwater Report", Technical report on groundwater management in the Mediterranean and the Water Framework Directive", EUWI, Brussels,.

Famiglietti et al. (2011), "Satellites measure recent rates of groundwater depletion in California's Central Valley", *Geophysical Research Letters*, Vol. 38(3).

Fanning, J. (2012), "Direct testimony of Jasper Fanning, Manager of the Upper Republican Natural Resources District, in the Supreme Court of the United States, State of Kansas (Plaintiff) *vs*. State of Nebraska and State of Colorado (Defendants)", No. 126.

Fornés, J., et al. (2007), "Legal aspects of groundwater ownership in Spain", *Water International*, Vol. 32(4), pp. 676–684.

Fuentes, A. (2011), "Policies towards a sustainable use of water in Spain" OECD Economics Department Working Paper No. 840, OECD, Paris, France. http://www.oecd-ilibrary.org/economics/oecd-economics-department-working-papers_18151973

Gardũno, H. and S. Foster (2010), *Sustainable groundwater irrigation: approaches to reconciling demand with resources*, GW-MATE, Strategic Overview Series 4, The World Bank, Washington, DC.

Garrido, A. et al. (2006), "Groundwater irrigation and its implications for water policy in semiarid countries: the Spanish experience", *Hydrogeology Journal*, Vol. 14(3), pp. 340–349.

Giordano, M. and K. Villholth (eds.) (2007), *The Agricultural Groundwater Revolution: Opportunities and Threats to Development*, CABI, Oxford.

Grafton, R. Q., et al. (eds.) (2014), *Global Water: Issues and Insights*, Australia National University Press.

Groot, N. (2013) "Norm Groot: Projects address seawater intrusion", *The Californian*, 16 December 2013. http://www.thecalifornian.com/story/news/2013/12/16/norm-groot-projects-address-seawater-intrusion/4036377/

Hanak, E. et al. (2014), "Funding sustainable groundwater management in California", 3 April 2014, UC Davis Center for Watershed Sciences, California Water Blog. http://californiawaterblog.com/2014/04/03/funding-sustainable-groundwater-management-in-california/.

Harter, T. and H.E. Dahlke (2014), "Out of sight but not out of mind: California refocuses on groundwater", *California Agriculture*, Vol. 68(3), pp. 54–55.

Hashimoto, J. (2013), "Japanese Municipalities' Efforts to Conserve Groundwater", *Japan for Sustainability Newsletter* 133, September, http://www.japanfs.org/en/news/archives/news_id034270.html.

Hathaway, D. (2011), "Transboundary groundwater policy: Developing approaches in the Western and Southwestern United States", *Journal of the American Water Resources Association*, 47(1), 103–113.

Hendricks, N. P., and J.M. Peterson (2012). Fixed Effects Estimation of the Intensive and Extensive Margins of Irrigation Water Demand. Journal of Agricultural and Resource Economics, Vol. 37(12), pp. 1–19.

Hernández-Mora, N. et al. (2010), "Groundwater in the Southern Member States of the European Union: an assessment of current knowledge and future prospects. Country report for Spain", European Academies Science Advisory Council (EASAC), Halle.

Howitt, R., et al. (2014), "Economic Analysis of the 2014 Drought for California Agriculture", Center for Watershed Sciences, University of California Davis, CA.

Iziquierdo, M. R., E. López-Gunn, and R.M. Llamas (2011), "Emergence and persistence of Groundwater User Collectives in Spain", *Ground Water*. http://www.rac.es/ficheros/doc/00830.pdf

Joshi, S.R. (2005), "Comparison of groundwater rights in the United States: Lessons for Texas", Master Thesis, Civil Engineering, Texas Tech University, TX.

Kemper, K.E. (2007), "Instruments and institutions for groundwater management" in Giordano and Villholth (eds.), *The Agricultural Groundwater Revolution: Opportunities and Threats to Development*, CABI, Oxford.

Kuwayama, Y. and N. Brozović (2012), "Analytical hydrologic models and the design of policy instruments for groundwater-quality management", *Hydrogeology Journal*, Vol. 20(5), pp. 957–972.

Lee, L. (2014), "Rural conservation ensures cities' water supplies, benefits landowners", tx | H2O, Summer 2014, Texas Water Resources Institute, Texas A&M University, College Station, TX. http://twri.tamu.edu/publications/txh2o/summer-2014/protect-our-land-protect-our-water/

Llamas, R. (2004), "Water and Ethics. Use of groundwater", UNESCO Series on Water and Ethics, Essay, UNESCO, Paris.

Llamas, R. (1975), "Noneconomic motivations in ground-water use: Hydroschizophrenia", *Ground Water*, Vol. 13, pp. 296-300.

Llamas, R.M. and A. Garrido (2007), "Lessons from intensive groundwater use in Spain: Economic and social benefits and conflicts", in Giordano, M., and K.G. Villholth (eds.), *The Agricultural Groundwater Revolution: Opportunities and Threats to Development*, CABI Oxford.

Lopez-Gunn, E. and M.R. Llamas (2008), "Re-thinking water scarcity: Can science and technology solve the global water crisis?", *Natural Resources Forum*, Vol. 32(3), pp. 228–238. doi:10.1111/j.1477-8947.2008.00200.x

Lopez-Gunn et al. (2012a), "Taming the groundwater chaos", in de Stefano, L. and M.R. Llamas (eds.), *Water, Agriculture and the Environment in Spain: Can we square the circle?*, CRC Press, London.

Lopez-Gunn et al. (2012b), "Implications of the modernization of irrigation systems", in de Stefano, L. and M.R. Llamas (eds.), *Water, Agriculture and the Environment in Spain: Can we square the circle*?, CRC Press, London.

Maliva, R.G. (2014), "Groundwater banking: opportunities and management challenges", *Water Policy*, Vol. 16(1), pp. 144-156.

Martin de Santa Olalla, F. et al. (1999), "Improvement of irrigation management towards the sustainable use of groundwater in Castilla-La Mancha, Spain", *Agricultural Water Management*, Vol. 40 (2-3), pp. 195-205.

Martín de Santa Olalla, F., A. Calera, and A. Domínguez (2003), "Monitoring irrigation water use by combining Irrigation Advisory Service, and remotely sensed data with a geographic information system", *Agricultural Water Management*, Vol. 61 (2), pp. 111-124.

McCabe, D. (2013), "N-CORPE to help meet river flow needs", *Nebraska Farmer*, Vol. 14.

McCarl, B. et al. (1999), "Limiting pumping from the Edwards Aquifer: An economic investigation of proposals, water markets, and spring flow guarantees", *Water Resources Research*, Vol. 35(4), pp. 1257-1268.

McCarthy, E. (2014), Biologist speaks out against deep-water wells. Farm Focus of Atlantic Canada, 8/05/2014, http://www.atlanticfarmfocus.ca/Agriculture/2014-05-08/article-3717650/Biologist-speaks-out-against-deep-water-wells/1;

McCook Gazette (2010), "$3 million in penalties for water meter cheats," 25 June, http://www.mccookgazette.com/story/1645392.html.

McKusick, V. (2002), "State of Kansas v. State of Nebraska and State of Colorado: Joint Motion of the States for Entry of Proposed Consent Judgement and Approval and Adoption of Final Settlement Stipulation", Supreme Court of the United States, 2002.

Mechlem, K. (2012), "Legal and Institutional Frameworks", Thematic Paper 6, Groundwater Governance: A Global Framework for Action, GEF, FAO, UNESCO-IHP, IAH, and the World Bank.

Montginoul, M. et al. (2014), "Controlling groundwater exploitation through economic instruments: Current practices, challenges, and innovative approaches", in S. Wheeler(ed.), *Integrated Groundwater Management*.

Morris, B.L et al. (2003), "Groundwater and its Susceptibility to Degradation: A Global Assessment of the Problem and Options for Management", United Nations Environment Program, Nairobi.

Nebraska DNR and TPNRD (Nebraska Department of Natural Resources and Upper Republican Twin Platte Natural Resources District) (2013), "Integrated Management Plan", Lincoln, Nebraska.

Nebraksa DNR and MRNRD (Nebraska Department of Natural Resources and Middle Republican Natural Resources District) (2010), "Integrated Management Plan", Lincoln, Nebraska.

Nebraska DNR and URNRD (Nebraska Department of Natural Resources and Upper Republican Natural Resources District) (2010), "Integrated Management Plan", Lincoln, Nebraska,

OECD (2015a), "The Role of Public Policies in Promoting Adaptation in Agriculture" [COM/TAD/CA/ENV/EPOC(2014)13/FINAL], OECD, Paris, http://www.oecd.org/officialdocuments/publicdisplaydocumentpdf/?cote=COM/TAD/CA/ENV/EPOC(2014)13/FINAL&docLanguage=En.

OECD (2015b), *Water in cities: ensuring sustainable futures*, OECD Studies on Water, OECD Publishing, Paris. http://dx.doi.org/10.1787/9789264230149-en.

OECD (2015c), *Water Resources Allocation: Sharing Risks and Opportunities*, OECD Studies on Water, OECD Publishing, Paris. http://dx.doi.org/10.1787/9789264229631-en

OECD (2014a), *Climate Change, Water and Agriculture: Towards Resilient Systems*, OECD Studies on Water, OECD Publishing, Paris. http://dx.doi.org/10.1787/9789264209138-en.

OECD (2014b), "Water resources allocation: Country profiles", OECD, Paris. http://www.oecd.org/fr/publications/water-resources-allocation-9789264229631-en.htm

OECD (2013), *Environment at a Glance 2013: OECD Indicators*, OECD Publishing, Paris. http://dx.doi.org/10.1787/9789264185715-en.

OECD (2013b), Making Water Reform Happen in Mexico, OECD Studies on Water, OECD Publishing, Paris. http://dx.doi.org/10.1787/9789264187894-en

OECD (2012a), A Framework for Financing Water Resources Management, OECD Studies on Water, OECD Publishing, Paris. http://dx.doi.org/10.1787/9789264179820-en

OECD (2012b), "Meeting the water coherence challenge", in OECD, *Meeting the Water Reform Challenge*, OECD Publishing, Paris. http://dx.doi.org/10.1787/9789264170001-7-en

OECD (2012c), *OECD Environmental Outlook to 2050: The Consequences of Inaction*, OECD Publishing, Paris. http://dx.doi.org/10.1787/9789264122246-en

OECD (2011), *Benefits of Investing in Water and Sanitation: An OECD Perspective*, OECD Studies on Water, OECD Publishing, Paris. DOI: http://dx.doi.org/10.1787/9789264100817-en

OECD (2010a), *Sustainable Management of Water Resources in Agriculture*, OECD, Paris, France. http://www.oecd-ilibrary.org/environment/oecd-studies-on-water_22245081

OECD (2010b), "OECD Member Country Questionnaire Responses on Agricultural Water Resource Management", Unpublished, OECD, Paris. Available at: http://www.oecd.org/agriculture/44763686.pdf

Palazzo, A. and Brozović, N., (2014), "The role of groundwater trading in spatial water management", *Agricultural Water Management,* Vol. 145, pp. 50-60. DOI: 10.1016/j.agwat.2014.03.004.

Peck, J.C. (2007), "Groundwater management in the High Plains Aquifer in the USA: Legal problems and innovations", In in Giordano and Villholth (eds.), *The Agricultural Groundwater Revolution: Opportunities and Threats to Development*, CABI, Oxford.

Ribeiro, L. and L. Veiga da Cunha (2010), "Groundwater in the Southern Member States of the European Union: an assessment of current knowledge and future prospects", Country report for Portugal, European Academies Science Advisory Council (EASAC), Brussels.

Ronen, D., S. Sorek and J. Gilron (2012), "Rationales behind irrationality of decision making in groundwater quality management", *Ground Water*, Vol. 50(1), pp. 27–36. doi:10.1111/j.1745-6584.2011.00823.x

Saak, A.E. and J.M. Peterson (2007), "Groundwater use under incomplete information", *Journal of Environmental Economics and Management*, Vol. 54(2), pp. 214–228. doi:10.1016/j.jeem.2006.12.005

Scanlon, B.R. et al. (2012), "Groundwater depletion and sustainability of irrigation in the US High Plains and Central Valley", *Proceedings of the National Academy of Science*, doi:10.1073/pnas.1200311109.

Schaible, G. and Aillery, M. (2012), "Water Conservation in Irrigated Agriculture: Trends and Challenges in the Face of Emerging Demands", United States Department of Agriculture Economic Research Service, Economic Information Bulletin (EIB-99), Washington DC.

Schoengold, K and D. Zilberman (2007), "The economics of water, irrigation, and development", *Handbook of Agricultural Economics*, Vol. 3(06).

Scott, C.A. and T. Shah (2004), "Groundwater overdraft reduction through agricultural energy policy: insights from India and Mexico", *International Journal of Water Resources Development*, Vol. 20(2), pp. 149–164.

Shah, T (2008), *Taming the Anarchy: Groundwater governance in South Asia, Resources for the Future*, Washington DC, and International Water Management Institute, Colombo, Sri Lanka.

Shah, T. et al. (2008), "Groundwater governance through electricity supply management: Assessing an innovative intervention in Gujarat, western India", *Agricultural Water Management*, Vol. 95(11), pp. 1233–1242.

Sharatt, S. (2014),"PEI potato growers want deep well permits", *Farm focus of Atlantic Canada*, 16 January, http://www.atlanticfarmfocus.ca/Agriculture/2014-01-16/article-3579226/PEI-potato-growers-want-deep-well-permits/1.

Struzik, E. (2013). *Underground Intelligence: The need to map, monitor, and manage Canada's groundwater resources in an era of drought and climate change*, Munk School of Global Affairs, University of Toronto, Toronto, http://powi.ca/wp-content/uploads/2013/05/POWIUndergroundIntelligence-Struzik-June25.pdf, accessed July 2013

Sommer, L. (2014), "As Water Prices Soar, Some Profit From California's Drought", 23 June, KQED Science, KQED Public Radio, Sacramento, CA. http://blogs.kqed.org/science/audio/some-california-farmers-fallow-fields-others-sell-water-for-big-profits/#comments

Sophocleous, M. (2012), "Conserving and extending the useful life of the largest aquifer in North America: the future of the High Plains/Ogallala aquifer", *Ground Water*, Vol. 50(6), pp. 831–9.

Sophocleous, M. (2010), "Review: groundwater management practices, challenges, and innovations in the High Plains aquifer, USA — lessons and recommended actions", *Hydrogeology Journal*, Vol. 18, pp. 559–575.

Turkish Ministry of Environment and Forestry (2009), "Water and DSI", Ankara.

UNECE (United Nations Economic Commission for Europe) (2011), "Second Assessment of Transboundary Rivers, Lakes and Groundwaters" Executive Summary, UNECE, Geneva.

US EPA (US Environmental Protection Agency) (2014), "Aquifer recharge and Aquifer Storage and Recovery", webpage, EPA, Washington, DC, http://water.epa.gov/type/groundwater/uic/aquiferrecharge.cfm.

Walker, A. (2014), "Debate over potato irrigation rages on in PEI", *Troymedia*, 24 August, http://www.troymedia.com/2014/08/24/debate-over-potato-irrigation-rages-on-in-pei/.

Water in the West. (2013), "The Water Energy Nexus: Literature Review", Institute for Sustainable Futures, Stanford University, Palo Alto, CA.

Wester, P., J. Hoogesteger, and L. Vincent (2009), "Local IWRM organizations for groundwater regulation: The experiences of the Aquifer Management Councils (COTAS) in Guanajuato, Mexico", *Natural Resources Forum*, Vol. 33(1), 29–38. doi:10.1111/j.1477-8947.2009.01206.x

Wichelns, D. (2010), "Agricultural Water Pricing: United States", in OECD, *Sustainable Management of Water Resources in Agriculture*, OECD Publishing, Paris. DOI: http://dx.doi.org/10.1787/9789264083578-16-en.

Worthington, B (2014), "Mapping and monitoring groundwater in Victoria", *Vic Country Hour*, ABC net Australia, April, www.abc.net.au/news/2014-04-29/victoria-groundwater/5417308.

Wright, T. (2014). "Agriculture federation wades into deep wells debate", *The Guardian*, 14 March, http://www.theguardian.pe.ca/News/Local/2014-03-14/article-3650183/Agriculture-federation-wades-into-deep-wells-debate/1.

Wright, W. et al. (2012), "Agricultural Groundwater Policy During Drought: A Spatially Differentiated Approach for the Flint River Basin", Selected paper for the 2012 annual AAEA meeting.

Yarr, K. (2014a),"Groundwater politics: the high-capacity wells debate", CBC News, 13 June, http://www.cbc.ca/news/canada/prince-edward-island/groundwater-politics-the-high-capacity-wells-debate-1.2674811.

Yarr (2014b), "Water for irrigation: the high-capacity well debate", CBC News, 18 June, http://www.cbc.ca/news/canada/prince-edward-island/water-for-irrigation-1.2674865.

Young, R.K. (2014), "Tradable permit systems for a spatially heterogeneous externality: A microparameter approach", Master's thesis, University of Illinois at Urbana-Champaign.

Annex 4.A1

Deriving regional indicators of groundwater characteristics and management

The large number of incomplete variables within the 20 regions makes it difficult to conduct any meaningful statistical analysis to link characteristics to policies. Instead, a set of characteristics indicators was developed based on responses representing the likelihood of groundwater stress, and a set of policy indicators were derived to stand for the scope of policy responses to control groundwater use.

The variables are defined below. They are computed by translating categorical responses in the questionnaire as discrete numbers, either with dummy variables or by using a categorical response as shown in parenthesis. For example, a region with an arid climate will be counted as 4 in the variable AgClim. When multiple categories are used, a simple average is taken.

Table 4.A1.1. Definitions of the variables used to represent the characteristics of the groundwater system

Question	Defined indicator
Agro-climatic zone	**AgClim** = Arid(4), Semi-arid(3), Temperate(2), Humid(1)
Climate change prospective (2030-50)	**Cchum** = Drier(2), Wetter(1), No significant change in precipitation(0)
	Cctemp = Hotter(2), Cooler(1), No significant change in temperature(0)
	Ccflo = More frequent floods
	Ccdry = More frequent drought
Type of aquifer	**Aqtype** = Confined(2), Unconfined(1), Mixed(1.5)
	AqGeol = Sand and gravel(1), Sandstone(2), Karst(3), Volcanic(4), Basement(5)
Surface irrigation	**SWav** = Surface water is available or not available?
	SWuse = Surface water is used or not used for irrigation?
	SWimp = The dominant source of water(1), Used conjunctively(2) Rarely used?(3)
Volume of groundwater irrigation	**Gwuse** = Volume of groundwater irrigation (year, estimate, unit)
	Gwevoluse = Diminishing(0.5), Steady(1), Increasing(1.5)
Other uses of groundwater	
Domestic	**UseDom** = Minor(1), Major(2)
	UseDomevol = Diminishing(0.5), Steady(1), Increasing(1.5)
Industry	**UseInd** = Minor(1), Major(2)
	UseIndevol = Diminishing(0.5), Steady(1), Increasing(1.5)
Mining	**UseMin** = Minor(1), Major(2)
	UseMinevol = Diminishing(0.5), Steady(1), Increasing(1.5)
Energy	**UseEn** = Minor(1), Major(2)
	UseEnevol = Diminishing(0.5), Steady(1), Increasing(1.5)
Other	**UseOth** = Minor(1), Major(2)
	UseOthevol = Diminishing(0.5), Steady(1), Increasing(1.5)

Table 4.A1.1. Definitions of the variables used to represent the characteristics of the groundwater system (*cont.*)

Question	Defined indicator
Externalities	
Pumping lift/cost increase	**ExtPump** = Minor(1), Major(2)
	ExtPumpevol = Growing(1.5), Steady(1), Reducing(0.5)
Well yield reduction	**Extyred** = Minor(1), Major(2)
	Extyredevol = Growing(1.5), Steady(1), Reducing(0.5)
Stream depletion	**ExtStrDep** = Minor(1), Major(2)
	ExtStrDepevol = Growing(1.5), Steady(1), Reducing(0.5)
Vegetative stress	**ExtVstr** = Minor(1), Major(2)
	ExtVstrevol = Growing(1.5), Steady(1), Reducing(0.5)
Ingress of polluted water	**ExtPollu** = Minor(1), Major(2)
	ExtPolluevol = Growing(1.5), Steady(1), Reducing(0.5)
Salinity	**ExtSal** = Minor(1), Major(2)
	ExtSalevol = Growing(1.5), Steady(1), Reducing(0.5)
Aquifer compaction	**ExtAqcomp** = Minor(1), Major(2)
	ExtAqcompevol = Growing(1.5), Steady(1), Reducing(0.5)
Land subsidence	**ExtLsub** = Minor(1), Major(2)
	ExtLsubevol = Growing(1.5), Steady(1), Reducing(0.5)
Other	**ExtOth** = Minor(1), Major(2)
	ExtOthevol = Growing(1.5), Steady(1), Reducing(0.5)

Table 4.A1.2. Definitions of the variables used to represent the characteristics of the groundwater system

Question	Defined indicator
Groundwater ownership	**Gwown** = Private(2), Public(1), Both(1.5), Neither(0)
Groundwater entitlement characteristics	**Entchar** = Permanent(1), Temporary(2) **EntLand** = Linked to land rights **EntTrfr** = Transferable
Beneficiaries of entitlement:	**EntBenInd** = Individuals **EntBenComp** = Companies
Groundwater entitlement allocation system:	**EntAlloc**= Absolute ownership(4), Reasonable(3), Correlative(2), Prior appropriation(1)
Groundwater management plans:	**ManPlan** = Mandated(2), Voluntary(1), None(0)
Co-ordination with surface water management:	**ManConjUse** = Systematic(2), Partial(1), Limited(0.5), None(0)
Regulations on wells:	**RegWappr** = Approval of new wells **RegWspace** = Accounting for well space restriction **RegWeia** = With environmental impact assessment **RegGW** = Goundwater withdrawals restriction
Regulations on irrigated land:	**RegIrarea** = Regulations on irrigated areas **Regexplrar** = Regulation on the expansion of irrigated **IrrLdBo** = Irrigated land buyout

Table 4.A1.2. Definitions of the variables used to represent the characteristics of the groundwater system (*cont.*)

Question	Defined indicator
Mandated metering or monitoring system for groundwater	**MetAg** = Agricultural users **Metfreq** = Report frequency (month=0.5 and year=1) **MetEnf** = Are these measures enforced (yes=1)
Economic instruments to regulate quantity: Pricing	**PriceAg** = Are there charges on pumped water? In agriculture **PricCR** = If so, are they based on cost recovery? In agriculture **PricExt** = Do they account for environmental externalities? in agriculture **PricScar** = Do they account for the scarcity value of water? in agriculture
Groundwater markets:	**Gmtempent** = Are temporary entitlements marketable? **GMLTent** = Are long term entitlements marketable? **GMentBo** = Are water entitlement buy-out possible?
Collective management schemes	**Cmvol** = Voluntary(self-regulation) **Cmreg** = Framed by regulations **Cmsta** = Mandated by states **Cmwua** = Water user associations **Cmdist** = District or community based **Cmoth** = Other
Agriculture water conservation programs	**AgWsub** = Subsidies **AgWloan** = Loans **AgWCond** = Conditional payments **AgWpen** = Penalty **AgWOth** = Other
Irrigation programs	**IrrSub** = Generic irrigation subsidies **IrrEffSub** = Irrigation subsidies focusing on efficiency **IrrEffLoan** = Loans for irrigation efficiency improvements
Energy programs	**ElecTax** = Electricity tax **ElecSub** = Electricity subsidies **ElecOth** = Other energy supporting programs (diesel, natural gas)
Land policies with implications on groundwater use	**Lpolzon** = Zoning with restrictions on groundwater use **Lpolreg** = Regional allocation system groundwater priority for other uses **LpolRUcoop** = Urban rural co-operation
Watershed conservation programs affecting groundwater use	**Wconszon** = Exclusion zone for conservation area **Wconslim** = Limits of groundwater use close to protected areas **WconsentBO** = Acquired groundwater entitlement for water conservation\
Climate change adaptation programs affecting agricultural groundwater use	**CCpolRD** = Investment in agric R&D **CCpolGw** = Investment in groundwater R&D **Ccpolinfr** = Water infrastructure invt **CCpolGwdata** = Groundwater modelling and data development
Aquifer recharge programs	**ASR** = Aquifer storage and recovery programs **Gwbank** = Groundwater banking **InfPond** = Infiltration ponds
Programs supporting the development alternative water supplies	**Winfrexp** = Surface water reservoir expansion **Desal** = Desalination **RecycW** = Recycled water **RwHarvest** = Rainwater harvesting
Agricultural income support policies	**SubBiof** = Biofuel production support
Drought insurance programs	**SubDrInsCrop** = Government subsidiesd plans for field crops **SubDrIns** = Government based insurance

Table 4.A1.3. Formulas used to compute the indicators

Indicator	Formula
Status and characterisation of agricultural groundwater systems	
IndClim	AgClim * (1+ Cchum + Cctemp + Ccflo + Ccdry)
IndAq	Aqtype + AqGeol
IndSW	SWav * (1 + SWuse) + SWimp
IndGWuse	Gwuse
IndOtherUse	(UseDom*UseDomevol) + (Uselnd*Uselndevol) + (UseMin*UseMinevol) + (UseEn* UseEnevol) + (UseOth*UseOthevol)
IndExt	(ExtPump * ExtPumpevol) + (Extyred *Extyredevol) + (ExtStrDep*ExtStrDepevol) +(ExtVstr*ExtVstrevol) + (ExtPollu*ExtPolluevol) + (ExtSal*ExtSalevol) + (ExtAqcomp *ExtAqcompevol) + (ExtLsub*ExtLsubevol) + (ExtOth*ExtOthevol)
Management and policy approaches	
IndEntFTO	Gwown *(1+ Entchar + EntLand + EntTrfr + EntBenlnd + EntBenComp) + EntAlloc
IndMan	ManPlan + ManConjUse
IndRegWell	RegWappr * (1+ RegWspace + RegWeia)
IndRegGW	IndRegWell + RegGW* (1+ Reglrarea + Regexplrar) + IndMet + IndMan
IndMet	MetAg * (1+ Metfreq)* (1+ MetEnf)
IndEconGW	PriceAg * (1+ PricCR + PricExt+ PricScar) + (Gmtempent*(1+ WconsentBO + GMentBo)) + (GMLTent *(1+ WconsentBO + GMentBo)) + EntTrfr + IrrLdBo + GMentBo
IndCMGW	Cmvol + Cmreg + Cmsta + Cmwua + Cmdist + Cmoth
IndOthControl	AgWloan + AgWCond + AgWpen + AgWOth + IrrEffSub + ElecTax + ElecOth + Lpolzon + Lpolreg + LpolRUcoop + Wconszon + Wconslim + CCpolRD + CCpolGw + Ccpolinfr + CCpolGwdata
IndSupply	ASR + Gwbank + InfPond + Winfrexp + Desal + RecycW + RwHarvest
IndOthConso	SubBiof + SubDrlnsCrop + SubDrlns + IrrSub + AgWsub + ElecSub

Note: These indicators are meant to help illustrate the existence and type of constraints and management approaches used by regions. Their construction is not meant to provide any precise measure, but rather project the data obtained in the questionnaire for comparison purposes.

Annex 4.A2

Results of the regional indicator analysis

The thirteen indicators were compared together and via pairwise correlation. The indicators were standardised by subtracting the sample average and dividing it by the standard deviation. The goal of the exercise is not to derive meaningful absolute values of stress and policy coverage, but rather to enable a comparison of the relative importance of the types of constraints and approaches used among regions. The computed indicators used for analysis are shown in Tables **4.A2.1** and **4.A2.2**, with the correlation in Table **4.A2.3**. Figures **4.A2.1** and **4.A2.2** provide an overview of the results with regions grouped by continent.

Table 4.A2.1. **Indicators on the main resource and use characteristics for the fifteen regions**

	IndClim	IndAq	IndSW	IndGWuse	IndOtherUse	IndExt
AusMDB	14		3			
DenWJ	10	2.5	5		1.5	
FraDV	10	4.5	4		3	
FraNB	12	1.5	5	0.22	2	2
IsrWG	21	5	4		6	12.5
ItaC	18	1	3	0.013	5	1
JapKKH	7.5	5.5	4	0.205	2	2.5
JapKS	7.5	2.5	4	0.442	2	2
JapNS	7.5	2.5	4	0.103	2	1
KorJVI	10	5.5	3	0.319044	4.5	6
MAA	15	2	5	12.22	5.5	1
MexRL	18	3	2		4.5	19
MPW	21	1.5	4	24.7	9	8.5
NHPA	15	1	4	7.72	7.5	3
NldG	6	2	5	0.015	2	0
NldL	6	2.5	5		2	0
NldNB	6		5		6	
PorTRO	14	2	4	0.597	2	
SHPA	15	2	4.5	11.16	9	8
SpaMOc	12	3	3.5	0.0912	2	14

Source: Derived from the OECD questionnaire (see Annex 4.A for computations).

Table 4.A2.2. Indicators on policy approaches

	IndEntFTO	IndRegGW	IndEconGW	IndCMGW	IndSupply	IndOthControl	IndOthConso
AusMDB	8	4.5	6	0	1		
DenWJ	1.5	4					
FraDV	6	4	2	3	0.25	10	4
FraNB	7	5	2	4	0.25	10	3
IsrWG	8	12	2	2	2	7	1
ItaC						2	1
JapKKH	9	3	0	0	0.33	0	0
JapKS	9	6	0	0	0	0	0
JapNS	9	6	0	0	0	0	0
KorJVI	11.25	12.5	3	3	1.083	6	2
MAA	13	2	0	0	0	5	3
MexRL	12	7	7	3	1.75	5	2
MPW	12.5	5.5	2	1	0.917	5	3
NHPA	11	11.7	6	2	0.667	5	3
NldG	3	8	0	0	0	6	2
NldL	3	8	0	0	0	6	2
NldNB	3	8	0	0	0	6	2
PorTRO	13	8	3	2	0.25	12	4
SHPA	12	5	3	0	0	5	3
SpaMOc	9	13	3	3		12	1

Source: Derived from the OECD questionnaire (see Annex 4.A for computations).

Table 4.A2.3. Pairwise correlation across selected indicators

	IndEntFTO	IndRegGW	IndEconGW	IndCMGW	IndSupply	IndOthConso	IndOthCont
IndClim	**0.62**	0.11	**0.58**	0.34	**0.70**	0.27	0.15
IndSW	**-0.51**	-0.29	**-0.73**	-0.39	**-0.71**	0.22	0.11
IndAq	-0.07	0.16	-0.11	0.17	**0.40**	-0.29	-0.01
IndOtherUse	**0.46**	0.10	0.36	-0.10	0.32	0.33	-0.10
IndExt	**0.42**	0.38	**0.69**	**0.54**	**0.85**	0.06	**0.40**
IndGWuse	**0.54**	-0.30	0.16	-0.23	0.32	**0.48**	-0.03

Source: Derived from the OECD questionnaire.

Figure 4.A2.1. Comparison of standardised indicators of groundwater characteristics in the 20 regions

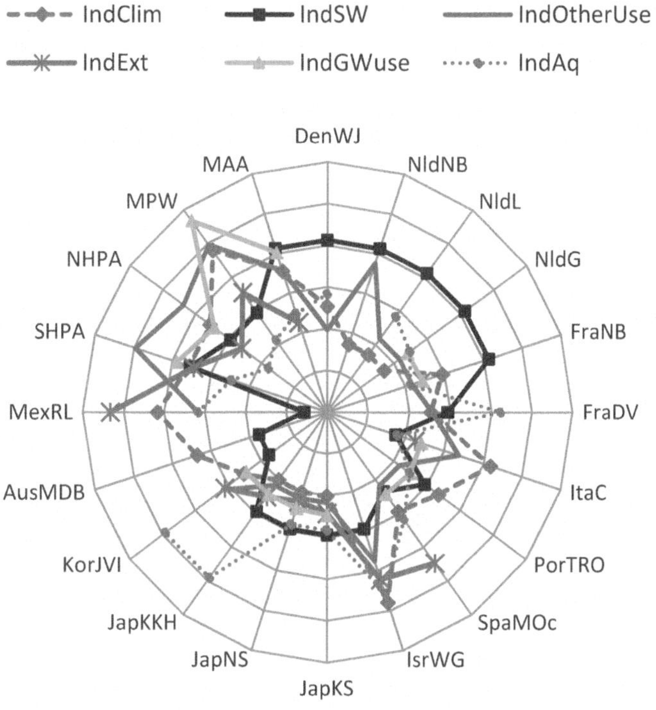

Each line represents a variable of interest.The closer they are to the centre of the circle, the lower the indicator value. For instance, the Western Jutland region (DenWJ) presents relatively high surface water availability (IndSW relatively distant from the centre), but lower indicators of other uses (IndOtherUse) and climate (IndClim), with lines closer to the centre, as it is not facing too much competition or climatic stress compared to other regions in the set.

Note: AusMDB: Murray-Darling Basin; DenWJ: Western Jutland, FraNB: Nappe de la Beauce, Fra DV: Departement de la Vienne; IsrWG: Western Galilee; ItaC: Campania (Ufita); JapKH: Kikuchi Heiya; JapKS: Kinaguwa Seigu; Jap NS: Noubiheiya Seigu; KorJvl; Jeju volcanic Island; MexRL: Region Laguna; NldG: Gelderlan and Overjissel ; NldL: Limburg NldNB; North Brabant; IsrJRB: Jordan river Basin; MexRL: Region Laguna; PorTRO: Tejo e Ribeiras do Oeste; SpaMOC: Mancha Occidental; MAA: Mississippi Alluvial Aquifer; MPW: Mountain and Pacific West; NHPA: Northern High Plains Aquifer; SHPA: Southern High Plains Aquifer;

Source: Derived from results from the 2014 OECD questionnaire.

Figure 4.A2.2. Comparison of groundwater management and policy indicators in the 20 regions

Higher panel: The seven indicators

Lower left panel: Main demand-side approaches Lower right panel: Other approaches

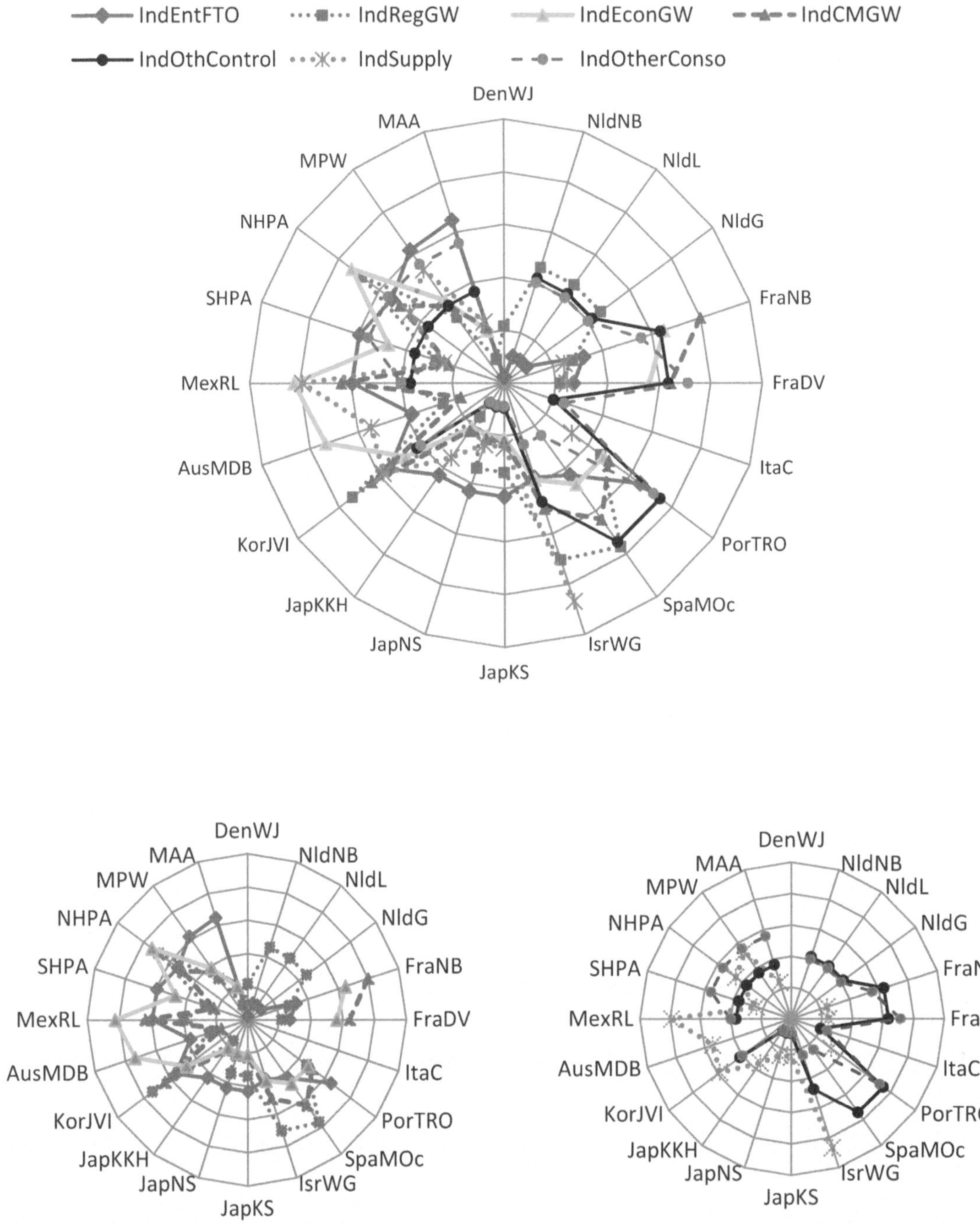

Lines closer to the centers represent low policy indicators, meaning that there is a relative lower (or less intensive) use of the specific approach compared to the group.The SpaMOc region for instance has a relatively high index for groundwater regulations,and a moderate index of economic instruments relative to the others.

Note: AusMDB: Murray-Darling Basin; DenWJ: Western Jutland, FraNB: Nappe de la Beauce, Fra DV: Departement de la Vienne; IsrWG: Western Galilee; ItaC: Campania (Ufita); JapKH: Kikuchi Heiya; JapKS: Kinaguwa Seigu; Jap NS: Noubiheiya Seigu; KorJvl; Jeju volcanic Island; MexRL: Region Laguna; NldG: Gelderlan and Overjissel ; NldL: Limburg NldNB; North Brabant; IsrJRB: Jordan river Basin; MexRL: Region Laguna; PorTRO: Tejo e Ribeiras do Oeste; SpaMOC: Mancha Occidental; MAA: Mississippi Alluvial Aquifer; MPW: Mountain and Pacific West; NHPA: Northern High Plains Aquifer; SHPA: Southern High Plains Aquifer;.

Source: Derived from results from the 2014 OECD questionnaire.

Chapter 5.

Towards adaptive groundwater management in agriculture

This chapter combines lessons from Chapter 3 and from past policy successes and failures to identify a package of recommendations for sustainable groundwater management. It then evaluates whether OECD policies fit this framework and concludes on the need for improvements in the context of the expected growing importance of groundwater management for agriculture under climate change.

The statistical data for Israel are supplied by and under the responsibility of the relevant Israeli authorities. The use of such data by the OECD is without prejudice to the status of the Golan Heights, East Jerusalem and Israeli settlements in the West Bank under the terms of international law.

Key messages

Building on lessons from the economic literature (Chapter 3) and complementary evidence of policy failures and successes, this chapter presents a three-part package of recommendations for sustainable groundwater management in agriculture, consisting of:

- Six general conditions for successful agricultural groundwater policies that outline the need to:
  - Build and maintain sufficient knowledge of groundwater resources and use.
  - Use surface- and groundwater management conjunctively (together), where relevant.
  - Target groundwater use directly rather than via indirect instruments where possible.
  - Enhance the enforcement of regulatory measures before moving to other approaches.
  - Prioritise demand–side approaches that affect users' incentives.
  - Avoid non-water related price distorting policy measures, such as subsidies for water intensive crops that could affect groundwater use.
- For regions using groundwater intensively, the use of a "tripod" combination of regulatory, economic and collective management approaches is recommended. In particular, groundwater entitlements systems should remain the core of groundwater management on which to add other instruments.
- Additional measures that increase agricultural water productivity and emerging new recharge mechanisms that provide complementary tools to traditional policy approaches, of relevance especially for highly stressed groundwater regions.

This framework should be adapted to locally-specific agriculture groundwater systems. This may call for the division of management into functional sub-units.

A significant share of responding OECD countries or regions with intensive agricultural groundwater use do not apply the recommended approaches. In particular, there seems to be insufficient knowledge on groundwater resources and use. Most OECD countries or regions in the survey sample have entitlement systems as their core groundwater management approach, but fewer countries and regions include economic or collective management approaches in their policy package.

For a much broader set of OECD countries than considered in this report, investing in information collection and analysis on groundwater resources, and configuring a balanced set of management instruments are necessary to achieve sustainable use of groundwater and continued agricultural productivity growth in a changing climate. Groundwater in many OECD regions is expected to play an increasing role in agriculture in the context of climate change and increasing variability in surface hydrological systems. Improving information systems on groundwater resources and flows should be the priority for all countries using or planning to use groundwater for irrigation.

Drawing recommendations from successes, failures, and lessons learned

As noted throughout this report, intensive groundwater pumping can generate significant externalities affecting agriculture and other users. This calls for management solutions. Chapter 2 shows there are diverse groundwater systems and challenges. Chapter 3 notes that management is needed to tackle externalities and points out that multiple policy options can help address these issues, acknowledging that specific responses will need to be tailored to hydrogeological and environmental conditions. Chapter 4 provides an overview of management approaches in OECD countries that vary widely, with weak evidence of a constraint-response correspondence.

The objective of this chapter is to move from towards a set of potential policy improvements to respond to the challenges of agricultural management. This chapter builds on a body of evidence that combines the conclusions of Chapter 3, the lessons learned from existing policy experiences (reported in Chapter 4 and in the literature), the results from the 2014 OECD questionnaire, and inputs from consultations with academic and institutional experts.[1] By identifying policy gaps and areas of concerns, this chapter aims to point to areas for which the current set of policies is likely to be insufficient and/or needs to be reconsidered, as well as areas where new policies could contribute to improved water management, accounting for current *and* future agro-climatic conditions.

Limited reporting evidence, but multiple lessons from the literature

There is only limited empirical reported evidence on successes and failures. Only four of the 20 responding OECD countries (Austria, Czech Republic, Slovenia and the Netherlands - national and regional) and five of the 20 regions (the Jeju volcanic island region in Korea and the four Spanish regions) report conducting regular evaluations of groundwater management. As part of the EU Water Framework Directive, EU member countries are required to provide reports on the state of implementation and the qualitative and quantitative status of groundwater management bodies. But they do not necessarily assess the specific policy approaches.

Still, the conclusions of Chapter 3 and a number of research publications provide some references for the analysis of successes and failures in addressing the challenges associated with agricultural groundwater use. These lessons put together can help identify key recommendations for improved groundwater management where systems are under constraint.

The relevance of these lessons and the derived recommendations clearly depend on the challenges to be tackled. A number of aquifers in OECD countries are either not used or used in a relatively sustainable manner, e.g. Northern European countries, and therefore generally do not need advanced agricultural groundwater management. But reported cases of lowering water tables, ecosystem damages, saline intrusion, stream depletion, and land subsidence observed in some of the most important OECD agricultural regions provide evidence of highly damaging external effects that call for policy responses.

Conditions for an effective groundwater management

With the above-mentioned caveats, six general conditions for sustainable management of groundwater use in agriculture are identified below.

a) *Build and maintain sufficient knowledge of groundwater resource and use.* Lack of information on groundwater resources is bound to lead to an inability to identify and adequately treat groundwater problems. Information collection is costly and needs to respond to a demand, but insufficient investment in groundwater information and data will prevent effective management.

– Depletion and externalities cannot be managed without any information on groundwater resources.[2] Even the core question of whether intervention is needed relies on a sufficient level of available information.

– The absence of information can also prevent considering the value of groundwater to policy makers, and the opportunity it could represent in low using communities (Foster et al, 2009).

– Groundwater cannot play its adaptation role with no information on groundwater reserves and the implication of its uses (Sophocleous, 2012). The large uncertainty on recharge prevents any meaningful groundwater management response (Crosbie et al., 2013). Aquifer storage and recovery in particular requires an advanced knowledge of the hydrogeology to avoid costly mistakes (e.g. see for instance Blood and Splagat, 2013).

– Furthermore, information can trigger cost-effective voluntary conservation measures. In the United States, for instance, advances in mapping and monitoring groundwater have led to innovative conservation measures, persuading ranchers and landowners to pull their water resources together in Utah or leading to a self-imposed voluntary 20% reduction of withdrawals by farmers in Northwest Kansas (Struzik, 2013).[3]

b) *Use surface and groundwater management conjunctively where relevant.* Ignoring existing connections with surface water in management can be very costly. Such behaviour has been found detrimental to aquifers and surface water bodies. The example of Tablas de Damiel National Park in Spain, shown in Box 5.1, is telling. Conjunctive water management is known to have multiple economic benefits (Schoengold and Zilberman, 2007) and is bound to be an effective way of reducing vulnerability to climate change, increasing water and food security (Taylor et al., 2012). At the same time, the OECD (1989) recommendations state the need to combine surface and groundwater management where relevant; in multiple areas, links between aquifer and surface water are not significant, and irrigation systems may rely only on groundwater. In such case, there would be no need to consider this option (e.g. GWP, 2012).

c) *Favour instruments that directly target groundwater use, where possible*. Indirect instruments (on land, energy, agriculture) may be more difficult to implement, less effective and/or leading to unwanted consequences. For instance, controlling new wells may be less costly and burdensome than monitoring their actual use by thousands of pumpers (Liu et al., 2014). More generally, legal, economic and institutional approaches that focus on water use for agriculture will be more effective than those that focus on linked inputs or outputs.

d) Prioritising demand-side approaches that affect groundwater users' incentives. Demand-side approaches address the root of the economic problem, while supply-side responses delay or avoid the constraints. Supply-side approaches should only follow, if necessary, demand requirements once demand side approaches have been fully implemented (Lopez-Gunn et al., 2012a).

e) *Enhance the enforcement of regulatory measures before moving to other approaches.* Partial enforcement of regulatory measures is bound to result in mismanagement of groundwater. In particular, the presence of hundreds, thousands or even millions of illegal wells is bound to eliminate the usefulness of additional measures (OECD, 2010b). Illegal withdrawals can also threaten ecological systems (Dionisio and Mario, 2014) and impede surface water management objectives and their associated instruments, such as water markets (Zetland and Weikhart, 2013).

f) Last but not least, *avoid non-water related price distorting policy measures that could affect groundwater use*. Subsidies and price distorting measures that undermine the marginal cost of groundwater and therefore move farmers outside of the optimal path will result in inefficient and potentially costly and undesired outcomes. For instance, in the Guajanato region of Mexico, the marginal cost of pumping is maintained artificially much lower than it should be, discouraging farmers from conserving the resources (Scott and Shah, 2004). By changing the structure of incentives, these policies bias decisions towards groundwater use (against alternatives, including surface water) and disincentive reductions in consumption with short and long run implications.

Box 5.1. Redressing groundwater externalities: The Tablas de Daimiel National Park in Spain

The Tablas de Damiel National Park, Upper Guadiana Basin in Spain has undergone a dramatic turn of events with the rapid intensive development of groundwater irrigation in the Western Mancha Aquifer. This park, established in 1973, is known for its wetland and is recognised as a UNESCO Man and the Biosphere Reserve. During the 1970s and 1980s, significant increases in groundwater irrigation lead to a diversion of water from the Western Mancha away from the Park. From 1974 to 1984, withdrawals increased from 400 hm^3/year to 500hm^3/year when the renewable level was believed to be around 260-300hm3/year. The irrigated area tripled in the same period. This phenomenon was encouraged by dry seasons, good prices, and related agricultural subsidies.

Such intensive use resulted in drops of the water table that reached 40 to 50m in some areas. Farmers dug more and deeper wells, increasingly competing with each other, and the wetland dried up. In 1994, the Guadiana Basin River Authority declared the area under "groundwater overexploitation". This triggered a series of regulatory measures: forbidding new wells, forcing the formation of water user associations, a strict reduction of the water quotas. However, the measures were not completely enforced and groundwater abstractions started to increase again at the end of the 1990s. In 2008, a Special Plan for the Upper Guadiana (SPUG) was introduced to address the 50% over allocation of groundwater resources.

This plan incorporated a range of regulatory, economic and collective actions with the objective of reaching "good status" under the EU Water Framework Directive, notably to stabilise use to a maximum level of 200hm^3/yr. The central measure of the plan was the purchase of water rights by the SPUC for large cereal farmers to restore and protect the environmental assets of the National Park. Seventy per cent of purchased rights were given to the Park and 30% to small productive farmers who focussed on higher value activities like vine and vegetables. This mechanism triggered a shift in groundwater use and agricultural activities away from cereals, encouraged vines to be accounted for, and led to increased incentives for the enforcement of water rights.

In 2011, thanks in part to good precipitation, the park increased its flooded area from 0 to 2000 ha and groundwater levels increased by an average of 17m. Still, the public cost of the program - estimated at 5000 M EUR for 2008-2027 - has been high and the program did not address the incentive structure of farmers pumping groundwater.

Source: Lopez-Gunn et al. (2012b).

Proposed policy package for sustainable management: A "tripod" combination

Moving to actual instruments, some overarching elements stand out from the collected evidence, while others are more difficult to assess. Table 5.1 provides the contour of a proposed groundwater package, building on a combination of three core (demand-side) approaches and two (agronomic and supply-side) additional elements to address the main concerns associated with intensive groundwater use. The proposed policy package is gradual with a differentiation based on groundwater stress level: the first set of measures relates to all groundwater using regions, the second to those under current or expected stress due to intensive use, and the third to very high stress regions. Correspondence to the six general conditions outlined above is highlighted in Table 5.1. While general, the framework should be adapted to locally specific constraints. The following paragraphs provide supporting evidence for the main management elements.

First, as noted in Chapter 3, there is no single superior instrument; multiple analyses and past experience strongly support the use of a *combination of approaches*.[4] Regulatory, collective and economic approaches, rather than each standing on its own, should be considered as a "tripod" of complementary levers on which to build groundwater management (Meinzen-Dick, 2007; Mechlem,

2012). In their review of groundwater irrigation policies, Garduño and Foster (2010) call for the design and use of pragmatic four tier approaches combining i) administrative measures (regulations and charges), ii) community involvement and self-regulation, iii) financing supply and demand interventions, and iv) macro-intervention to constrain groundwater demand (agriculture policies, energy subsidies). Barraqué et al. (2010) report that "most experts now agree on the need to supplement regulations and voluntary agreements with general economic incentives, i.e. abstraction and pollution charges, in particular on farmers, the former to be paid by volumes abstracted". Esteban and Dinar (2013) conclude on the superiority of packaged sequenced policies instead of individual ones.

Table 5.1. Proposed management package for intensive groundwater use

Main approaches	Associated instruments	Conditions for implementation	
		Instrument-specific	Cross-cutting
Prerequisite for action (applicable to all groundwater using regions)			
Information and monitoring systems	Regular data collection and analysis on groundwater resources, use, and surface-groundwater interactions	Sufficient investment in groundwater monitoring, metering, training and education.	
Core management approaches (for regions under intensive use)			
Entitlement system on which to develop regulations	Groundwater use entitlements regulations on new wells, metering and periodical reporting	Direct regulations on groundwater (decoupled from land entitlements) Monitoring and enforcement of existing measures first	
Collective management approaches	Water user association or other groundwater collective scheme.	Overarching framing regulations with higher objectives Provide freedom of operation to collective decisions	Account for surface – ground water interactions Ensure enforcement Adapt to local conditions
Economic instruments	Trading groundwater entitlements Pricing water	Provide an enabling environment Consider responsiveness Avoid secondary measures	
Additional approaches (to use especially under high stress)			
Agronomic and technical changes	Technical and agronomic measures Moving towards high value and/or high nature crops per drop	Irrigation efficiency measures need to be associated with complement regulatory cap.	
Supply-side recycling measures	Groundwater recharge via banks or ASR, using recycled water. Rainwater harvesting and use of recycled water	Only to use in complement with a demand-side system, based on robust scientific analysis	

Lessons from countries' and regions' successful experiences also confirm these conclusions. Fuentes (2011) supports the use of economic incentives for water user associations to co-operate and enforce regulatory requirements on groundwater in Spain. Sophocleous (2012) reports that the relative successes that have allowed to slow down groundwater overdrafting in Kansas can be attributed to the multiple approaches that have been implemented, such as: the establishment of groundwater management districts, minimum stream flow regulations, metering and monitoring of resources, integrated resource planning, aquifer storage and recovery (ASR), a central water bank, and various water conservation programs. The local management approach that effectively slowed salinity intrusion in the Pajaro Valley, California combined multiple actors and actions, including supply-side to recharge, water charges, metering, changes of agricultural practices, education and information, and community-based demand restriction (Levy and Christian Smith, 2011).

Within the proposed tripod combination, the *groundwater entitlements* (permits or rights) remain the core measures on which to base locally customised regulatory schemes. As noted in Chapter 3, well metering and allocation of groundwater pumping entitlements can support new more finely targeted management instruments. Chapter 4 has demonstrated the large range of possible options: from environmental and land-related constraints at water wells, to conditions on uses that vary locally and over time. Among others, allowing for the purchase of entitlements by third parties for ecological or social equity purposes can be a key measure in reaching a social consensus on groundwater management (Lopez-Gunn et al., 2010).

Multiple sources also note the growing importance of *collective management schemes* as key for an effective management of groundwater bodies (Campana, 2014). Esteban and Dinar (2012) analyse agricultural water management options and find that collective responses always lead to higher welfare results for farmers and society. Indeed, self-regulation and management by user groups have been effective ways to conserve groundwater (FAO, 2011; Koundouri, 2004). Stephenson (1996) reported how the Upper Republican Nebraska River District helped control groundwater coming from the High Plains Aquifer. In Mexico, the Santo Domingo aquifer has largely benefited from a multi-stakeholder groundwater management initiative that included modern irrigation techniques, capacity building, and efforts to collect rainwater and recharge groundwater (OECD, 2013). Custodio (2010) notes the need to combine government actions and public management, regulations, stakeholder involvement, and co-responsibility. EASAC (2010)'s review of groundwater policies in Mediterranean countries concludes with a call that new groundwater management institutions move away from simple command-and control-elements to increasingly include self-governed management communities.

Regulatory requirements must play a complementary role to these schemes. Garduño and Foster (2010) note that self-regulation or collective management can be effective in specific settings, provided they are overseen by a local groundwater management agency. Lopez-Gunn and Martinez Cortina (2006) support a shift towards more self-regulation mechanisms for groundwater management, but with necessary backup regulatory regimes. Sophocleous (2012) reports that history has shown the limitation of voluntary mechanisms of halting on their own groundwater depletion; this mechanism works better with applied and enforced timelines and limits.

At the same time, collective management schemes should maintain a sufficient degree of autonomy in order to thrive. Forcing collective decisions among competing users and imposing management objectives and methods to groups will not generally result in win-win solutions. Cases in Spain and Mexico, among others, have shown that there can be limitations in restricting collective approaches. The emphasis should be on balancing framing regulations with sufficient autonomy for groups of users to manage groundwater.

Third, *economic instruments* can support some of the most efficient solutions to groundwater scarcity and depletion problems, provided they are implementable and not associated with high transaction costs. The diversity of instruments should help support locally specific considerations. As noted in Chapter 3, trading groundwater entitlements provides flexibility and leads to efficient allocation solutions, letting users set the price. But they cannot be adapted to all contexts and can be constrained, in particular, by entry and transaction costs (Koundouri, 2004; Garrido et al., 2012). Pricing (charges or taxes) can be an effective means to manage use if designed and used properly in contexts of elastic demand, but may face political challenges (Chapter 3 and OECD, 2009). Subsidies do not always provide the right incentive to conserve water. Other economic schemes acting on land, encompassing trading, or supporting conservation practices can be effective at a local scale if the above-listed solutions are not implementable.

Lastly, two evolving sets of practices appear promising in providing complimentary help especially when facing increasing groundwater pressures.

- *Agricultural choices and related technical options* — that may be induced from regulations or economic schemes — have a role to play, from more efficient irrigation practices to changes in agricultural cropping systems that are less water-demanding (Madramootoo, 2012). These may be supported via research and development, information campaigns, advisory services or extension, or by altering other support programs in their direction. In Mexico, Scott and Shah (2004) advocate to downsize pump capacity while increasing irrigation efficiency, and to shift to lower-water-demand crops to address long-term groundwater overdraft. Regions with increasingly scarce water resources are encouraged to move from "more crop (and jobs) per drop" to "more cash and nature per drop" (Lopez-Gunn et al., 2010; Howitt et al., 2014).
- *Alternative water recycling, transfer and recharge* mechanisms related to conjunctive management at a local scale, involving multi-stakeholders, show much promise if they remain cost-effective. They may be worth considering especially in situations of increasingly visible scarcity or externalities so long as they are built on sufficient information about the hydrogeology and that they support a larger regulatory framework.

This three-part package remains general but it should be emphasised (as noted in Chapter 2) that local specifications matter in the outcome of policies. As Brozović et al. (2006) conclude: "optimal policy should vary idiosyncratically across space and time" to reflect groundwater characteristics and the corresponding second-best instruments will be better if they account for the spatial distribution of wells, water demand, and hydrological parameters. Liu et al. (2014) further find that optimal decisions to dig new wells and/or to pump groundwater are dependent on the underlying spatial externalities of the resources. This may call for the division of management into sub-units.

Several recent reforms are indeed moving in this direction. In Kansas, the division of aquifers into subunits was one of the key management approaches for local decision (Sophocleous, 2012). The EU WFD is also based on that principle, with the definition of hundreds of thousands of groundwater management bodies for locally customised management solutions (Box 4.2). The analysis conducted in Chapter 4 for a small number of region suggests some partial differentiation of management systems based on constraints, at least when considering the level of groundwater stresses. Still, more and better information is bound to help better customise solutions to constraints.

Each of these recommendations will generally not be implemented in specific programs related to groundwater in agriculture, but rather embedded into broader water resource policies or water allocation reforms. As such, it is important to note that each of these 11 guiding elements — six framing conditions and five instruments — is consistent with the principles guiding general water resource allocation systems. OECD (2015) defined a set of 14 checks at the user and system levels to fulfil in order to achieve economically efficient, equitable and sustainable allocation principles. At the system level, requirements for accountability and information and knowledge fit with the proposed information requirement. Furthermore, effective monitoring and enforcement, assuring the system's interconnectivity (conjunctive surface-groundwater management), and policy coherence across sectors (no price distorting measures) are mentioned above. OECD (2015) recommendations on the definition and possible extension of entitlement mechanisms, the use of water charges and transfers as allocative mechanisms, or the capacity to store water also appear to be aligned with the suggestions on the instruments proposed here.

Lastly, while this is outside of the scope of this report, groundwater policies will only be properly applied if they are implemented in a well-functioning governance system. As noted in Box 5.2, there needs to be a good co-ordination across all levels of governance.

Box 5.2. From policy to implementation: What governance system for groundwater?

There are multiple institutional set-ups to manage groundwater (Mechlem, 2012). Wijnen et al. (2012) differentiate three levels in groundwater governance systems: policy, strategic, and the local governance. The local specificities of groundwater bodies can be managed by local authorities and tend to fall within a framework of regional or national legal and political actions, but there are also intermediate levels of administration. These variable levels may not always align their objectives, approaches or instruments, thereby creating multiple layers of a complex picture. Furthermore, if the social and political pressures for groundwater management may come from national, state, or local levels, in general, local groundwater organisations and institutions are often the ones that are developing, implementing, and enforcing management mechanisms. Lastly, the relationship with the management of surface water is a critical aspect, especially in regions where they are used conjunctively with groundwater.

A global effort associating the United Nations FAO, UNESCO, the Global Environmental Facility, the International Association of Hydrologists, and the World Bank has focused on determining what would constitute a good model of governance for groundwater. They first provided a diagnostics, then proposed a framework for action. This framework includes six pillars: 1) understanding the context; 2) creating a basis for governance; 3) building effective institutions; 4) making essential linkages; 5) redirecting finances; and 6) establishing a process of planning and management.

Source: Mechlem (2012), Wijnen et al. (2012), GEF, World Bank, FAO, UNESCO and IAH (2015). Groundwatergovernance.org.

Are these recommendations used in current policy frameworks of OECD countries?

How are OECD countries scoring on these 11 guiding elements? Responses from the questionnaire are used to provide a partial *aperçu* of their alignment with these criteria on the basis of 20 proxy variables (Table 5.2). To reflect some of the differences, national level policies are separated from those applied in the responding regions. Some of these variables, particularly those related to information, are defined based on the rate of response of countries to specific questions. While this rate may not reflect the status of information in all cases, the fact that the information is difficult to obtain is in itself an indicator of information unavailability and/or the effectiveness of groundwater information systems.

The national level results are separated into two categories, the first regrouping the top ten surveyed groundwater countries in agriculture, the second all 20 responding countries. Results vary regarding data and information collection, with 32% to 78% of responding countries fulfilling the recommended provisions (top part of the table) and an even wider range among higher groundwater-using countries. Most responding countries, and four of nine high users, meet the other general conditions. Few countries have applied the proposed tripod combination of policy instruments (lower part of the table): only seven of the 20 responding countries (35%) and three of the nine responding high users use regulatory, economic and collective management approaches simultaneously. Entitlement remains the core management approach in most responding countries. But other types of instruments are not as common. Interestingly, higher groundwater users are more likely to use collective management than economic approaches, in contrast with other countries. Alternative water supplies and R&D programs are also employed by over a third of responding OECD countries, but less in higher groundwater user countries.

For the regional level responses, the three segments distinguished at the top of Figure 4.9 — representing relative degrees of groundwater stresses — are separated in Table 5.2. Regions with moderate groundwater stress present the highest rates of fulfilment on most of the proposed elements and instruments for agriculture groundwater management. At least some of the five most stressed groundwater regions checked the main conditions. In contrast, a number of the least stressed agricultural groundwater regions did not comply with the list of conditions, with the exception of conjunctive management and the use of direct instruments. On the management side, highly stressed regions would benefit from reviewing their approaches, especially via more use of collective management schemes. Whether the relatively wide use of the proposed recommendations by the regions in the intermediate category contributes to their moderate stress status would require further data and analysis. Conversely, it is not unreasonable to consider that some stresses may be needed to induce comprehensive responses.

Taken together, these results suggest that some of the OECD countries and regions that face agriculture groundwater challenges have generally adopted partial or incomplete approaches to their management. The gathered evidence does not allow for a definitive judgment on the performance of varied choices, but existing evidence collected here and analysed in Chapter 3 would favour reconsidering the framework under which these systems are governed. This is especially true when taking into account climate change.

An increasing need for a more sustainable management of groundwater resources to face a changing climate

Throughout Chapters 3 and 4, the emphasis was largely around the intensive use of groundwater for agriculture irrigation. This means in particular that certain OECD countries are more directly concerned due to the challenges they currently face. But increasing knowledge about projections of the climate change impact on freshwater availability (Chapter 1) implies that a larger number of countries and regions are likely to be turning to groundwater pumping in the future. Indeed, in the questionnaire ten of the 15 analysed agricultural regions report they expect significant changes in groundwater with climate change. As Figure 5.1 shows, 19 of the 23 regions (83%) responding to this question expect higher temperatures, and 87% expect more frequent droughts. Semi-arid regions are also expected to be drier, and semi-humid to be wetter; floods will also likely be more frequent in 61% of the regions.

Table 5.2. Countries and regions that implement the proposed groundwater management package[1]

	Indicator	Policies operating at the regional level			National level policies	
	Share of response rates or number based on other information sources	Groundwater stressed agricultural regions[2]	Moderately groundwater stressed agricultural regions[3]	Less stressed agricultural groundwater regions[4]	Largest using countries[5]	All
General conditions						
Data and information collection	Reported data on total groundwater resources	0%	25%	43%	22%	41%
	Reported data on groundwater recharge	60%	50%	14%	67%	37%
	Reported data on total groundwater use	n/a	n/a	n/a	78%	74%
	Reported data on agricultural groundwater use	40%	100%	57%	100%	78%
	Reported data on agriculture wells in use	40%	50%	43%	n/a	n/a
	Monitoring wells and/or reports	20%	50%	29%	n/a	n/a
	Reporting Information programs	n/a	n/a	n/a	33%	35%
	Groundwater data development	0%	0%	0%	33%	40%
Comprehensive approaches	Co-ordination with surface water management [6]	40%	50%	86%	44%	70%
Direct management approaches	More direct than indirect instruments	60%	62.5%	100%	44%	65%
Enforcement of regulations	No evidence of illegal wells/partial enforcement	n/a	n/a	n/a	67%	82%[7]
Accounting for farmers' incentives	Presence of economic approaches [8]	60%	37.5%	0%	55%	70%
Price distorting measures	Absence of potentially distortionary policies [9]	n.a.	n.a.	n.a.	78%	70%
Management instruments						
Combining the three approaches	Presence of the three types of approaches	20%	25%	0%	33%	35%
Groundwater entitlement as core approach	Entitlements for groundwater	60%	75%	100%	67%	70%
Collective management at the local level	Water user association or district based initiatives	20%	37.5%	14%	55%	30%
Economic instruments	Charges or water trading	60%	37.5%	0%	33%	50%
Agricultural choices and technologies	R&D programs	n/a	n/a	n/a	33%	55%
Alternative water recycling, banking and ASR	Presence of the relevant programs	20%	25%	14%	22%	35%

Notes: 1. Only those countries that responded to the relevant questions of the OECD 2014 questionnaire are included here. 2. AusMDB, MPW, SHPA, IsrJvB, MexRL, 3. MAA, NHPA, FraDV, ItaC, JapNS, KorJVI, PorRTO; SpaMOc 4.FraNB,DenWJ, JapKS, JapKH, NldG, NldL, NldNB. 5. Responding countries within the top 10 in groundwater use: AUS, ESP, ITA, JPN, KOR, MEX, PRT, TUR, USA. 6. Partial or systematic. 7. Share based on countries listed in the report, out of 34 OECD countries rather than responding ones. 8. Direct or indirect. 9. Generic irrigation subsidies, income support programs.

Source: 2014 OECD questionnaire.

Figure 5.1. Expected climatic evolutions in responding regions by climate (number of responses)

Left panel: increased water stresses, Right panel: increase rainwater supply

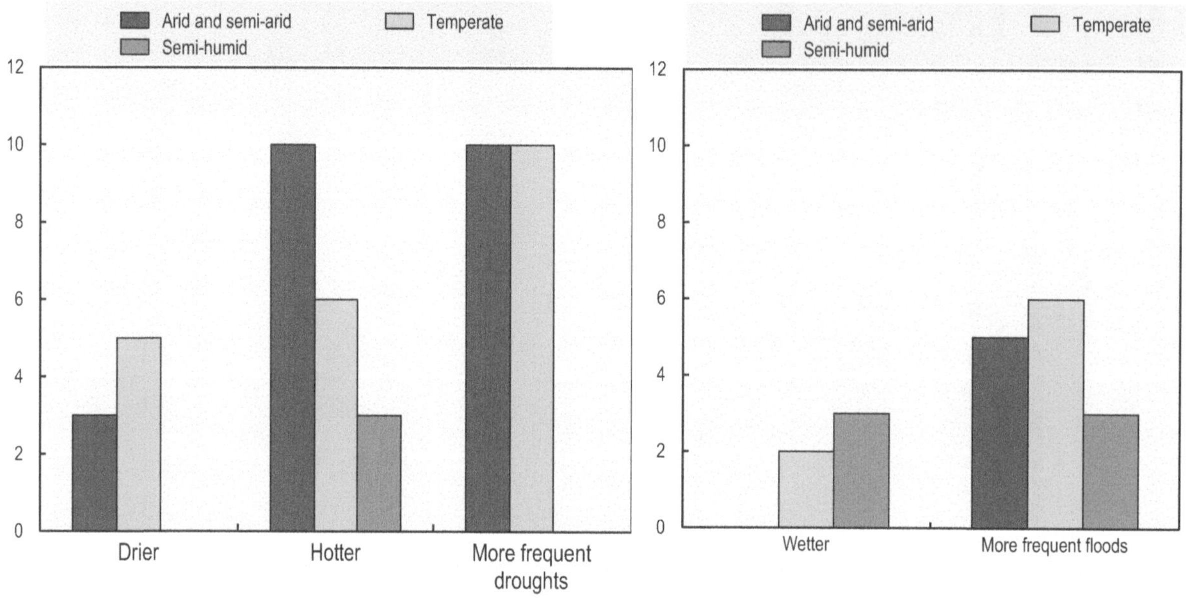

Source: 2014 OECD questionnaire.

There are two potential consequences of climate changing conditions on agriculture groundwater policies: i) current management instruments may need to be adapted to changing conditions in agricultural groundwater using regions; and ii) new instruments, including those reviewed in this report, may become necessary in agricultural regions and countries that do not significantly use groundwater today. In the first case, some large groundwater-using regions may face more scarcity and extreme risk situations. This may induce the development of emergency adaptation solutions and, in some cases, shifting to management solutions that pertain to non-renewable resources management, such as those used in mining or oil extraction. In the second case, regions that currently experience relatively lower groundwater stress may need to move towards additional and alternative instruments to cope with greater groundwater use and competition. This may include a shift towards a more advanced combination of instruments, including economic measures and potentially recycling and aquifer recharge programs. At the same time, some of the humid regions may need to manage additional precipitation by rethinking their conjunctive water management approaches, to reduce recharge and increase use, e.g. with more pumping, drainage, and less surface water use for irrigation, to mitigate the potential risks of water logging and salinity (Chapter 2).

Both types of shifts necessitate a move towards a better set of management practices. This chapter helped identify some of the key aspects of groundwater management on which to build locally-specific strategies (Table 5.1). Of these, two primordial items appear to be prominently missing in OECD countries, and/or necessary to ensure that groundwater is better used in the future.

First, there is a clear need for better information and data collection, analysis and dissemination for all stakeholders involved, not only in OECD countries but at the global level. New discoveries of groundwater sources in Africa and under the seafloor, as well as unprecedented published estimates of groundwater depletion in Northern India and western United States show there is progress, at least in some regions. But these examples also demonstrate that the global assessment of groundwater resources and use is far from complete. In the survey conducted for this report, 22%% of the 20 responding OECD countries were not able to provide national estimates of agricultural groundwater use. Over half of these countries did not report groundwater resources or recharge rates. Information may be disseminated in a number of organisations without inter-agency communication in some countries, but many respondents report not having any data on the matter.

There is also a lack of *ex ante* and (more so) *ex post* analysis on sets of instruments that can or are used to manage groundwater resources in agriculture. More empirical research and empirical meta-analyses are needed to better understand the correspondence between local characteristics, management solutions, and measurable results.

Information and knowledge are even more critical when considering potential climate impacts. Even if groundwater quantity and uses are not a concern for several OECD countries and regions today, they may become so in the future. Anticipating the availability and need of surface *and* groundwater, and how to shift strategies and potentially rely on groundwater for agriculture irrigation, cannot be done without knowing the status and dynamics of resource and uses, or without a clear understanding of how such challenges can be addressed.

This lack of information can be due to insufficient investment in monitoring and data, to a lack of co-ordination among actors, or to the fact that groundwater is highly localised, but also potentially due to other constraints. Groundwater often remains outside the political agenda, in part because of its invisible character. Even in water related institutions and networks, groundwater is often perceived as a "secondary" issue. At the international level, groundwater remains a low priority issue despite its growing importance. For instance, Jarvis (2013) notes that the Millennium Development Goals, even if considering water scarcity, do not explicitly consider groundwater.

Second, in light of climate change projections, the current use of policy instruments would need to be reviewed. If externality-inducing intensive groundwater use is generally not driven by supporting policies in OECD countries, incomplete policies that are badly enforced, and or relatively rigid in their implementation are likely to prevent the sustainable exploitation of groundwater for agriculture in the future. Surface and groundwater bodies remain subject to separate management regimes, when an increased emphasis on conjunctive use is needed. Countries should also exploit the potential promises of innovative local collective mechanisms, that combined with national or regional regulatory and economic policies seem to be among the most successful in tackling critical groundwater scarcity challenges. Public policy interventions, if deemed necessary, should prioritise locally adaptable demand-side approaches before considering moving to water supply investment to ensure farmer engagement in the future.

The overall objective should be to transform groundwater resources from being considered as only a productive input for agriculture to being valued as a long-term, climate-insulated reservoir that needs to be sustainably managed. If well managed, groundwater can and should act as a powerful adaptation option, a natural insurance mechanism, and not just a component of freshwater resource supplies.

Notes

1. See Chapter 1 for details.
2. In California, the lack of access to data on pumping prevented "any effective regional management plans for groundwater" in the past (Howitt et al., 2014).
3. Information provision is a necessary but not sufficient condition to groundwater management; it cannot redress externalities. Saak and Peterson (2007) show that in the absence of other groundwater management schemes, it can lead to strategic increased extraction and negative results in some specific cases.
4. This is consistent with OECD (1989)'s council recommendation which explicitly recommended the use of legal, regulatory (permits) and economic instruments adapted to groundwater systems.

References

Barraqué, B. et al. (2010), "Groundwater in the Southern Member States of the European Union : an assessment of current knowledge and future prospects. Country report for France", European Academies Science Advisory Council (EASAC) Halle.

Blood, M.R. and E. Spagat (2013), "Millions spent on water-storage plan that leaks", *Associated Press*, 26 August 2013, http://news.yahoo.com/millions-spent-water-storage-plan-leaks-184059156.html, accessed August 2013.

Brozović, N. et al. (2006), "Optimal management of groundwater over space and time", in R.U. Goetz and D. Berga (eds.), *Frontiers in water resource economics*, Springer, New York.

Campana, M. (2014),"Groundwater Management: Quo Vadis?" in Fitch, E.J. and R.A. Engberg (eds.), "AWRA at 50: The Future of Water Resources in the United States", *Water Resources Research*, Vol. 16(1), pp. 26-28.

Crosbie, R.S. et al. (2013), "Potential climate change effects on groundwater recharge in the High Plains Aquifer, USA", *Water Resources Research*, Vol. 49, pp. 3936–3951.

Dionisio, P. and G. Mario (2014), "Insuring water: a practical risk management option in water-scarce and drought prone regions?" *Water Policy*, Vol. 16, pp. 244-263.

EASAC (European Academies Science Advisory Council)(2010), "Groundwater in the Southern Member States of the European Union", EASAC Policy Report 12, EASAC, Halle.

Esteban, E. and A. Dinar (2013), "Modeling sustainable groundwater management: packaging and sequencing of policy interventions", *Journal of Environmental Management*, Vol. 119, pp. 93–102.

Esteban, E. and A. Dinar (2012), "Cooperative Management of Groundwater Resources in the Presence of Environmental Externalities", *Environmental and Resource Economics*, Vol. 54(3), pp. 443–469.

FAO (United Nations Food and Agricultural Organization) (2011), *The State of the World's Land and Water Resources for Food and Agriculture: Managing systems at risk*, Food and Agricultural Organization Publications, Rome.

Foster, S. et al. (2009), "Sustainable Groundwater Management Contributions to Policy Promotion", GW-MATE, Strategic Overview Series 1, The World Bank, Washington, DC.

Fuentes, A. (2011), "Policies Towards a Sustainable Use of Water in Spain", *OECD Economics Department Working Papers*, No. 840, OECD Publishing, Paris. DOI: http://dx.doi.org/10.1787/5kgj3l0ggczt-en

Gardũno, H. and S. Foster (2010), "Sustainable groundwater irrigation: approaches to reconciling demand with resources", GW-MATE, Strategic Overview Series 4, The World Bank, Washington, DC.

Garrido, A. et al. (2006), "Groundwater irrigation and its implications for water policy in semiarid countries: the Spanish experience", *Hydrogeology Journal*, Vol. 14(3), pp. 340–349.

Garrido, A. et al. (2012), "Water trading in Spain", in de Stefano and R.M. Llamas (eds.), *Water, Agriculture and the Environment in Spain: Can we square the circle?*, CRC Press, London.

GEF, World Bank, FAO, UNESCO and IAH (Global Environmental Facility, World Bank, UN FAO, UNESCO and International Association of Hydrogeologists) (2015), "Global groundwater governance: A framework for action", GEF and World Bank, Washington DC, FAO, Rome, UNESCO, Paris.

Giordano, M. (2009), "Global Groundwater? Issues and Solutions", *Annual Review of Environment and Resources*, Vol. 34(1), pp. 153–178.

GWP (Global Water Partnership)(2012), "Groundwater Resources and Irrigated Agriculture", GWP, Stockholm.

Howitt, R. et al. (2014), "Economic Analysis of the 2014 Drought for California Agriculture", Center for Watershed Sciences, University of California Davis, CA.

Jarvis, W.T. (2013), "Water Scarcity: Moving Beyond Indexes to Innovative Institutions", *Ground Water*, Vol. 51(5), pp. 663–9.

Koundouri, P. (2004), "Current Issues in the Economics of Groundwater Resource Management", *Journal of Economic Surveys*, Vol. 18(5), pp. 703–740.

Levy, M. and J.C. Smith (2011), "Groundwater Management in the Pajaro Valley", The Pacific Institute, Oakland, CA.

Liu, Z., et al. (2014), "Strategic Entry and Externalities in Groundwater Resources: Evidence from the Lab", Resource and Energy Economics.

Lopez-Gunn, E. and L. Martínez Cortina (2006), "Is self-regulation a myth? Case study on Spanish groundwater user associations and the role of higher-level authorities", *Hydrogeology Journal*, Vol. 14(3), pp. 361–379.

Lopez-Gunn et al. (2012a), "Taming the groundwater chaos", in de Stefano, L. and M.R. Llamas (eds.), *Water, Agriculture and the Environment in Spain: Can we square the circle?*, CRC Press, London.

Lopez-Gunn et al. (2012b), "Tablas de Daimiel National Park and groundwater conflicts", in de Stefano, L. and M.R. Llamas (eds.), *Water, Agriculture and the Environment in Spain: Can we square the circle?*, CRC Press, London.

Lopez-Gunn et al. (2010), "The impossible dream? The Upper Guadiana System: Aligning Changes in Ecological Systems with Changes in Social Systems", On the Water Front: Selections from Stockholm Water Week 2010-SIWI.

Madramootoo, C.A. (2012), "Sustainable Groundwater Use in Agriculture", *Irrigation and Drainage*, Vol. 61, pp. 26–33. doi:10.1002/ird.1658

Mechlem, K. (2012), "Legal and Institutional Frameworks", Thematic Paper 6, Groundwater Governance: A Global Framework for Action, GEF, FAO, UNESCO-IHP, IAH, and the World Bank, Washington DC.

Meinzen-Dick, R. (2007), "Beyond panaceas in water institutions", *Proceedings of the National Academy of Sciences*, Vol. 104(39), pp. 15200–15205.

OECD (2015), *Water Resources Allocation: Sharing Risks and Opportunities*, OECD Studies on Water, OECD Publishing, Paris. http://dx.doi.org/10.1787/9789264229631-en.

OECD (2013), *Making Water Reform Happen in Mexico*, OECD Studies on Water, OECD Publishing, Paris. http://dx.doi.org/10.1787/9789264187894-en.

OECD (2009), *Managing Water for All: An OECD Perspective on Pricing and Financing*, OECD Studies on Water, OECD Publishing, Paris. http://dx.doi.org/10.1787/9789264059498-en.

OECD (1989), "Recommendation of the Council on Water Resource Management Policies: Integration, Demand Management, and Groundwater Protection", OECD, Paris. http://acts.oecd.org/Instruments/ShowInstrumentView.aspx?InstrumentID=36&Lang=en&Book=False

Schoengold, K. and D. Zilberman (2007), "The economics of water, irrigation, and development", *Handbook of Agricultural Economics*, Vol. 3(06).

Scott, C.A. and T. Shah (2004), "Groundwater overdraft reduction through agricultural energy policy: insights from India and Mexico", *International Journal of Water Resources Development*, Vol. 20(2), pp. 149–164.

Stephenson, K. (1996), "Groundwater management in Nebraska: governing the commons through local resource districts", *Natural Resources Journal*, Vol. 505(1982).

Sophocleous, M. (2012), "Conserving and extending the useful life of the largest aquifer in North America: the future of the High Plains/Ogallala aquifer", *Ground Water*, Vol. 50(6), pp. 831–9.

Taylor, R.G. et al. (2012), "Ground water and climate change", *Nature Climate Change*, Vol. 3, pp. 322-329.

Wijnen, M. et al. (2012), "Managing the Invisible: Understanding and Improving Groundwater Governance", Water Partnership Program, Water Paper, The World Bank, Washington, DC.

Zetland, D. and H. Weikard (2013), "The institutional potential for water markets in the Tajo and Segura basins", Working Paper.

GLOSSARY

Aquifer Hydraulic continuous body of porous geological structure containing groundwater.

Groundwater Water contained in an aquifer matrix located beneath the surface in the saturated zone as opposed to free surface water bodies like streams, reservoirs, or lakes. It is constituted of the underground water that fully saturates all fissures and pores below the earth's surface.

Groundwater entitlement The entitlement to abstract and use groundwater from an aquifer system as defined in the relevant water plan or legislation. In a number of contexts, they may be referred to as "groundwater rights", abstraction licenses, or permits.

Groundwater entitlement allocation The following four doctrines have been used:

- ***Absolute ownership***: Doctrine also known as "Rule of capture or "English rule" in which the owner of a specific land also owns all water underneath and can pump water without limit (with specific exceptions, e.g., in case of malicious or wasteful use). There is no temporary permit required to use water.
- ***Correlative rights***: Doctrine under which groundwater is owned by the landowner with a requirement to share the aquifer. In other words, the landowner is considered a shareholder of an aquifer, and his rights are defined relative to others; groundwater is considered *de facto* a common property resource.
- ***Reasonable use:*** Doctrine under which the owner of the land is allowed to use groundwater without limit so long as it is for a reasonable purpose, with the definition of what constitutes reasonable purpose defined in legal texts. In practice, this implies that harm to neighbouring landowners, e.g. via overdrafting an aquifer, can result in liability.
- ***Prior appropriation:*** First in time, first in right" or "rule of priority" doctrine under which water is owned by the State and allocated based on seniority of use. It allows for the use of well permits and well assessments.

Land subsidence Lowering of the land surface induced by groundwater pumping. Drawing water in aquifers made of unconsolidated and porous geological structures, including sedimentary complexes, can result in significant and irreversible compaction of aquifers that in some cases result in the sinking of the land surface.

Pumping lift/ cost increase	Increase in the cost and lift needed to use groundwater due to intensive pumping.
Salinity	The increased concentration of salt in the water above a certain threshold based on the total concentration of dissolved solid.
Stream depletion	Depletion of the surface water level in streams induced by groundwater pumping close to waterways, rivers and/or lakes.
Vegetative stress	The effects of changes in depth to groundwater due to pumping on vegetation reliant on groundwater. These effects include changes in physiology, structure, and plant ecological dynamics, as observed particularly in arid regions.
Well yield reduction	A borehole is a hole in the earth crust for study or exploitation (well). Yield reduction is the diminution of the abstraction flow due to intensive exploitation.

ORGANISATION FOR ECONOMIC CO-OPERATION AND DEVELOPMENT

The OECD is a unique forum where governments work together to address the economic, social and environmental challenges of globalisation. The OECD is also at the forefront of efforts to understand and to help governments respond to new developments and concerns, such as corporate governance, the information economy and the challenges of an ageing population. The Organisation provides a setting where governments can compare policy experiences, seek answers to common problems, identify good practice and work to co-ordinate domestic and international policies.

The OECD member countries are: Australia, Austria, Belgium, Canada, Chile, the Czech Republic, Denmark, Estonia, Finland, France, Germany, Greece, Hungary, Iceland, Ireland, Israel, Italy, Japan, Korea, Luxembourg, Mexico, the Netherlands, New Zealand, Norway, Poland, Portugal, the Slovak Republic, Slovenia, Spain, Sweden, Switzerland, Turkey, the United Kingdom and the United States. The European Union takes part in the work of the OECD.

OECD Publishing disseminates widely the results of the Organisation's statistics gathering and research on economic, social and environmental issues, as well as the conventions, guidelines and standards agreed by its members.

OECD PUBLISHING, 2, rue André-Pascal, 75775 PARIS CEDEX 16
(51 2015 12 1 P) ISBN 978-92-64-23869-5 – 2015

IWA Publishing's authorised EU representative for General Product Safety Regulations is Diane D'Arras, 15 rue Duret, 75116 Paris, France, e-mail: safety@iwap.co.uk.

Printed and bound by CPI Group (UK) Ltd, Croydon, CR0 4YY
12/05/2026
02108865-0019